338
.274
0941
Ric

Richardson, John Bryning.
Metal mining. London, Allen Lane, 1974.
xvi, 207 p. illus., maps. (Industrial archaeology, 12)
Includes bibliography.

1. Nonferrous metal industries - Great Britain - History. 2. Mines and mineral resources - Great Britain. I. Title.

L-1887 M

I
A

Industrial Archaeology

A SERIES EDITED BY
L. T. C. ROLT

12

Metal Mining

Metal Mining

J. B. Richardson

Allen Lane

First published in 1974

Allen Lane
Penguin Books Ltd
17 Grosvenor Gardens, London SW1

ISBN 0 7139 0603 0

Printed in Great Britain by
Lowe & Brydone (Printers) Ltd,
Thetford, Norfolk

Contents

List of Illustrations

Line Drawings and Maps in the Text

Figure

Acknowledgements

I am grateful to Miss Regina Oblatt, Chief Librarian of the Institution of Mining and Metallurgy, without whose cooperation this book could never have been written, not only because of the constant flow of technical books but the added kindness of voluntarily checking obscure points by reference to the British Museum and other libraries. I particularly want to thank Mr H. L. Douch, Curator of the Truro Museum, Royal Institution of Cornwall, for providing papers from his archives and putting himself out to help and advise and Mr A. E. Truckell, M. B. E., Curator of the Burgh Museum of Dumfries, not only for information and advice but for providing excellent photographs of God's Treasure House in Scotland. I admire Miss Helen Christie and Miss Susan Gooday for deciphering my manuscript.

Preface

This book is the briefest of chronicles of what was a major British industry for thousands of years and at several periods of this island's history its *chief* industry. In order to keep it within the bounds of this series, the following chapters have been confined to mining the non-ferrous metals. Even so, the subject is a vast one, capable of almost infinite expansion. Nevertheless it is hoped that the reader will gain from the pages which follow a reasonably clear overall picture of an almost forgotten but vitally important facet of Britain's industrial past.

For although during the Industrial Revolution the production of tin, copper, lead and zinc was overshadowed by the upsurge in coal and iron output, it was essential to the success of this new machine age because every machine that moved on land or sea had to have some of these four metals. During that period tin was needed for bearing metals, bronze, pewter, britannia metal, solders, type metal, tinplate and coinage. Copper was essential for the then new electrical industry, bronze, brass, high duty bearings, bushes and coinage; by 1780 all the ships of the British Navy were copper-bottomed; lead was needed for pipes to convey water, for the covering of the newly introduced cables, for bearing metals, typemetal, solders, shot and paint; zinc, which came into prominence later, was beginning to be used in the young electrical industry in storage batteries.

It is not generally appreciated that these islands are more richly provided with mineral wealth than any country of equal size in the world. Apart from vast reserves of coal and iron ore, Nature has been generous in providing huge quantities of non-metallic minerals, from the best slates in the world and other materials used in building, road-making, concrete, pottery and glass to fluxes used in the iron and steel industry.

We were not quite so handsomely endowed, however, with the ores of the principal base metals, copper, lead, tin and zinc, which enter so largely into the needs of our modern civilization. Nevertheless mining the ores of the non-ferrous metals had been pursued from ancient times and during the middle of last century we were the world's largest producers of ores of copper, tin and lead and during the Industrial Revolution occurred the great booms in these metallic ores. Tin and lead continued to boom for several more years but every mineral deposit, being exploited, is a wasting asset and the only capital of a mining company is the deposit itself. When it is exhausted the company dies. Moreover the costly machinery required for mining and beneficiation is usually of such a special nature that it has little or no value for other purposes when the deposit is exhausted.

Throughout the Industrial Revolution metal mining was bedevilled by three factors, the cost book system of financing, the wide and rapid fluctuations in metal prices and the private ownership of mineral rights. In the nineteenth century taxation was negligible sometimes 7*d* in the £1 and after royalties were paid to the mineral rights owners the remainder of the profits were distributed. Up to the twentieth century metal mines were financed by the cost-book system which was tragic for the continuity of metal mining. The purser of the mine combined the functions of company secretary and accountant and had the company's business under his control and when a profit was made in any year it was divided amongst the adventurers. When a loss was made the adventurers were asked to meet it in proportion to their shareholdings. The adventurers thus had close control over the way their money was spent, but under this system no money was put to reserve and on the sudden closure of a mine, through a fall in metal prices or from other causes, there was no money for development in search of further ore or, in the case of the Cornish copper mines, to sink through the poor copper-tin and tin-copper zones to reach good tin ore. In Cornwall Dolcoath and Levant mines amongst others luckily had the money to continue and from rich copper mines became rich tin mines.

The price of gold and silver and the base metals is governed internationally. Except for gold, these prices fluctuated like a fever chart. This means that however efficient the management and however skilled and hard-working the miners, if the price fell sharply to the point that a British mine could no longer work at a profit it was forced to close down. So the deep Cornish tin mines, producing a raw material that had to be expensively processed to free the tinstone of impurities, could not compete with the cheaply won, clean, alluvial tinstone of Malaya,

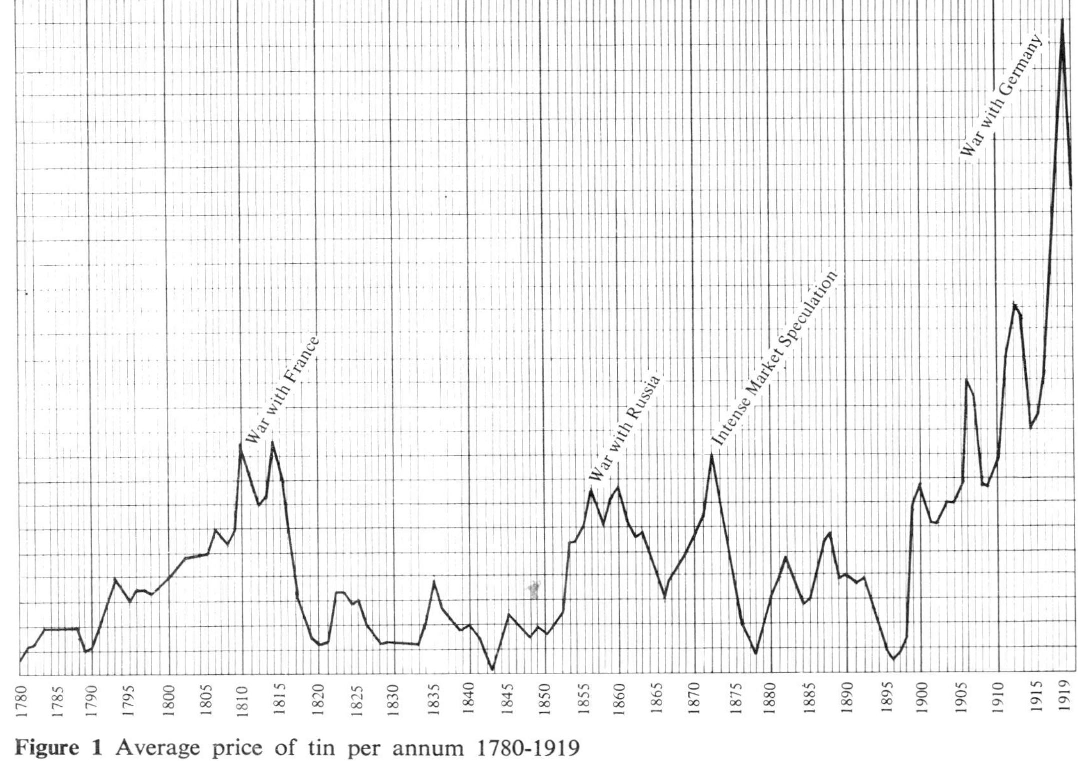

Figure 1 Average price of tin per annum 1780-1919

Thailand and Indonesia and all but two had to close down.

During the Industrial Revolution successive governments concerned themselves only with levying taxes and looking after the health and safety of persons employed in mines. They took no interest in profit snatching, allowing adventurers to pick the eyes out of a mine, thus ruining the future life of mineral deposits; nor were they concerned with the extortionate demands of mineral rights owners, who levied their royalties on the gross value of the concentrates irrespective of their net value even when a mine was working at a loss. Lead ores royalties were as much as a tenth of the selling price.

Many millions of words have been written by geologists, engineers and metallurgists devoted to the subject of seeking, finding and exploiting non-ferrous metallic ores and their reduction to metal in Great Britain. Many shelves in both the British Museum and the Science libraries are filled with books that tell the detailed history of this island's former mineral wealth. But although it is only four generations ago that this small country was the world's greatest producer of copper, tin and lead, so short are memories that the great urban majority knows little or nothing of this and cares less. Only the sparse population of the uplands of Scotland, Lakeland, the Pennines, wild Wales and mid-Cornwall are constantly reminded of this past because in these more remote areas lie the stone-arched entrances of soughs, horse-levels and other adits, the collars of innumerable old shafts, the miles of leats, those man-made water courses that fed water to the dressing floors, whose ruins are still to be seen. There are sometimes, too, the rusting remnants of machines, the ruins of engine houses or ancient smelters and, above all, the unsightly mounds of waste rock in which, even today, specimens of ore are still to be sought and found. All mining despoils the landscape and the cost of complete rehabilitation, even when possible, is so high that the feeling against any revival is strong.

During the eighteenth and nineteenth centuries mining and metallurgical practices caused widespread damage to the environment. The Industrial Revolution was based on the mining and burning of coal to provide steam, motive power and the production of iron. Earlier the iron ore had been reduced by charcoal, with the destruction of forests, causing the despoilation of the countryside; later coal produced the pollution of urban atmosphere by sulphurous fumes, creating a hatred of mining and metallurgical processes in the minds of the British public. It is doubtful if there is another country where the reaction against mining is as strong as it is in Britain which paid a heavy price for her leadership in exploiting coal and iron and, less importantly, the base metals: witness the devastation of Swansea Vale, the bare backbone of Cornwall and the hideous waste dumps in every British minefield.

The Industrial Revolution created an infinity of technical, commercial and social changes in this island, and if it cannot be precisely dated the names of such men as Abraham Darby, Watt and Henry Cort and their dates indicate its beginning.

In the non-ferrous mining and metallurgical field there were equally inventive men. John Taylor, who introduced the crushing rolls into Britain, was managing mines at the early age of nineteen and successfully improved the mining of tin, copper and silver-lead from Cornwall to Cumberland, was active in mid and north Wales and Derbyshire. He also founded a family firm of mining consultants, John Taylor & Sons, which still prospers. Smeaton of Eddystone Lighthouse fame was employed at Greenwich Hospital's smelter at Langley and by the Carron Company in Scotland. The Boulton and Watt partnership solved the dewatering of the Cornish copper mines by their powerful pumping engines. Before them came Newcomen working on the same problem and after them the great Cornish inventor Richard Trevithick. Percival Norton Johnson, F. R. S., a trained metallurgist, successfully revived the silver-lead industry in the Bere Alston peninsula, introducing many technical improvements and was the founder of the world-famed precious metal refiners, Johnson Matthey & Co.

Let it not be forgotton that when charcoal became hard to get and expensive, base-metal smelters began to experiment with pit coal many years before Abraham Darby, a Black Country man from Sedgley, succeeded in reducing iron ore with coke in 1709.

As far back as 1640 Sir Bevil Grenville tried to smelt tin ore with fossil fuel. John Coster was smelting both lead and copper successfully in 1687 with pit coal and although Viscount Grandison failed with his coal-fired reverberatory furnace in 1678, Dr Edward Wright the Quaker doctor succeeded in 1690. Dr John Percy, rightly called the father of British metallurgy, wrote in the preface to his *Treatise on Non-ferrous Metallurgy*, 'In no country are the operations of metallurgy conducted on so vast a scale as in Great Britain'. It is remarkable how many physicians became prominent metallurgists.

Pattinson and Parkes in mid-nineteenth century invented processes greatly increasing the extraction of silver from lead bullion. Robert

Stagg reduced the cost and improved the operation of the Pattinson process.

Such men improved the underground layout of mines, the dressing of ores, the transport and the management of mines and smelters and were the stock from which professional mining and metallurgical engineers grew.

CHAPTER ONE

Introductory

Four thousand years ago metal mining began in the British Isles.[1] At the close of the third millennium B.C. or the beginning of the second, Iberian prospectors in their small sailing boats, somewhat like the curraghs of south-west Ireland, crossed from Brittany to Cornwall to discover the stupendous eluvial tin deposits of Devon and Cornwall. They were short, wiry, long-headed, dark-complexioned men, and according to one Spanish historian were like the oldest race in Europe, the Basques.[2] Dr Glyn Daniel[3] considers they were associated with the megalithic religion which spread from Crete to Scandinavia.

Groups of these same adventurers crept up the coast of Wales, harbouring in river estuaries, and crossed to Ireland to find the biggest alluvial gold deposit in Europe in the rivers of County Wicklow, and copper ore galore at Avoca and other parts of Ireland, but no tin.[4]

Tin was then a strategic material, like oil or uranium ore today, because it was impossible to make weapons of bronze without it, and trade was active with the tin-hungry civilizations of the eastern Mediterranean basin. However indirect, there was undoubtedly contact between the Cretan bronze civilization and Britain, for on one of the uprights of Stonehenge is carved indistinctly a dagger of Mycenean type only discerned in 1953, and Irish gold ornaments have been found both in Crete and Mycenea.

The second millennium B.C. saw men in Cornwall and Devon with wooden tools mine the tin ore, and with wind and charcoal reduce it to metal. The Irish bronzesmiths had to have tin and Cornwall provided it. Cornwall was a stage in the journey from Ireland to Europe, and lay between the Irish and their customers. There is ample proof in the numerous relics of bronze and gold that have been unearthed that they crossed Cornwall from Hayle River to Marazion, for though bold and brave seamen they shrank from rounding stormy capes like Land's End where four seas meet. Numerous recent finds of golden objects in

England from Hull to Chatham and Flintshire to Dorset are all of Irish gold, hidden in panic and never recovered.

When the Phoenicians established a colony at Cadiz, ancient Gades, outside the Straits of Gibraltar, about 1100 B.C., and the Greeks had colonies at Marsilia (Marseilles) and Narbo (Narbonne) about 600 B.C., the tin trade became more highly organized and the metal collected at Ictis (St Michael's Mount) found its way to Brittany and then by donkey up the Loire valley and down the Rhône to Greek merchants. But the great metal traders of Phoenicia and Carthage got their tin at Corbilo[5], an entrepôt on rocky islands at the then unsilted mouth of the Loire to transport it by sea to Gades, thought by some historians to be the main distributing centre for British tin from 1000 to 200 B.C.

It is always sad to explode a long-established myth, but the late Sir Gavin de Beer[6] has shown that the Cassiterides were these rocky islets in the Loire estuary and not the Scilly Isles and St Michael's Mount. Much later, one of Caesar's generals, Publius Licinius Crassus, visited the Loire islets in about 56 B.C. and found tin smelting sites there. Moreover, no collection of Phoenician artefacts has ever been found in Cornwall, for besides purple cloth they bartered imperishable glass, pottery and beads for tin. In any case the identification of the Scilly Isles as the Cassiterides is nonsense because they are tinless.

The intrepid Greek sea-captain Pytheas, financed by Marseilles merchants, circumnavigated Britain and sailed to the coast of Norway about 325 B.C. His records are lost but he is quoted by later writers, who related that Pytheas found the Cornish 'intelligent and friendly because the tin traffic had brought them into contact with many different people'. He watched the tin miners at work and reported: 'They extract the tin from its bed by a cunning process. Having melted the tin and refined it they hammer it into knucklebone shapes and convey it to an island called Ictis. They wait until the ebb tide has drained the intervening firth and then transport whole loads of tin in waggons.'[7]

The bronze-using folk who invaded south-east Britain about 1900 B.C.[8]* are called Beaker folk from the shape of the pottery vessels from which they drank their barley beer. They were a hybrid race of battle-axe people from central Europe and the Beaker folk originally from Spain. They crossed from Holland and the Rhineland to make successive landings on the east coast of Britain. The Beaker folk included itinerant bronzesmiths who travelled the country mending and remelting bronze

*See page 180

weapons and tools; their craft was a secret handed down from father to son. These roundheaded, strongboned, muscular men mingled with the megalithic farmers, and many writers credit them with introducing the metal age to Britain. However, recent archaeological research has established that the Severn Valley bronze axes, daggers and halberts were of earlier Irish origin.[9]

The Wessex culture ended the early Bronze Age and flourished from 1500 to 1200 B.C. It is now considered that these powerful autocratic chieftains were natives, not invaders from Brittany. They gained their riches from cattle and the taxing of local inhabitants. In sumptuous graves their bones or ashes were buried in coffins of hollowed tree trunks, together with rich furniture, gold cups with handles, amber, triangular bronze daggers, some with gold-studded hilts and faience beads from Egypt.[10] They controlled the trade from the Baltic to central Europe and even to the Aegean.[11]

In the middle Bronze Age copper was mined at Alderley Edge in Cheshire, where the sandstone, impregnated with green copper carbonate, was easily reduced to metal. Probably copper ore was mined also in Anglesey, near Amlwch. Lead was mined in the Mendips and Derbyshire, and possibly in central Wales as well.

The main Celtic invasions arrived in south-east England from Champagne in waves from 500 to 450 B.C. The big, fairhaired Celts were a master race and were attracted by the mineral wealth of Britain. These iron-using conquerors introduced cheap durable weapons and tools for the felling of trees, and made possible the cultivation of richer, deeper soils so that the population increased. With their superior weapons they overran the British Isles, imposing on the natives their language, customs and religion. To dominate the tin trade they built forts like Chun Castle near St Just in Cornwall, with walls over fourteen feet thick enclosing a 170-foot diameter compound with eleven huts for smelting tin.[12]

About 150 B.C. the partly Romanized fierce Belgae began to enter south-east England, introducing wheeled vehicles and heavier ploughs. Before the Romans came Britain exported not only cattle, hunting dogs and slaves, but metals as well.

The knowledge the Romans had of the geography of the British Isles was scanty. In Gaul they had met the tribal chiefs wearing golden ornaments. Vercingetorix, who surrendered to Caesar at Alesia in 54 B.C., wore bracelets, a torque and a great round brooch of Irish gold.

So the Romans when they came were sadly disappointed in the lack of gold and silver, but, in their hunger for metal, they soon worked the base-metal deposits extensively. Britain's prosperity under Roman rule was due not to manufactures but to fertile fields and rich metal mines. The short easy haul to Gaul and the Rhine and the strong demand for lead in Rome led to a lively intercourse with the Continent.

Soon after the conquest began, the Romans set the local inhabitants to work. So quickly did trade follow that within six years Britain was exporting lead. Because the Mendips were within the area of early conquest they were the first to be exploited. Lead ingots from that area were counterstamped by the Second Agusta Legion and dated A.D. 49.[13] Possibly the soldiers were in charge of local labour, but if not the legionaries were quite capable of working the shallow diggings themselves as they were not only fighting men but expert in road-making, bridging and building forts.

Dated pigs prove that lead ore was mined in Flintshire by A.D. 74 and at Nidderdale in Yorkshire by A.D. 87. The Romans mined for lead in Flintshire, Derbyshire, Yorkshire, Shropshire and Alston Moor, as well as in the Mendips. How much military supervision was necessary, and whether there was universal forced labour is uncertain. The workings were mostly opencast on exposed outcrops. The miners were apparently amenable and did not cause much anxiety to the officials in charge. Where underground workings are known the entrances were open to the surface and the workers free to come and go. However, there is some archaeological evidence of barrack-like establishments in Derbyshire.[14] The turbulent Ordovices of North Wales, according to one writer, were tamed more by the attractions of city life than by force of arms, as the tribesmen brought the bullion as tribute to Chester and carried out the mining and smelting in their hills largely unsupervised. In the more distant parts of the province production was carried out under strict military supervision, but in the settled parts lead mining was sometimes handed over to private Roman concerns, notably the Socii Lutudariences of Lutudarum, that is Chesterfield in Derbyshire.[15] At the mines near Minsterley in Shropshire the 20th Legion took a hand in organising mining, according to a countermarked pig dated A.D. 195 found at Châlon-sur-Saône. The military certainly supervised the lead mining at Alston Moor, not very far from Hadrian's Wall. Details of all the other places where lead was mined in England and Wales, the pigs found, dated and undated, the abbreviated inscriptions and the translations thereof would fill a small book.

For the Romans, lead and the silver extracted from it was the most important British product. Pliny the Elder, in his *Natural History*, refers to lead most laboriously dug up in Spain ... 'but in Britain so abundantly in the upper crust of the earth that the law forbade more than a certain quantity should be mined'.[16] This was an early case of protection from foreign, British, competition, as Roman notables had financial interests in the Spanish mines. However in the third and fourth centuries Britain took over from Spain the precedence she had enjoyed since the end of the Second Punic War.

The Roman need for hard-wearing bronze for military and domestic uses encouraged the mining of copper ores. At Llamymynech in North Shropshire there were galleries and caves in which the miners lived. There is similar evidence at the copper mines on the Great Orme. Objects of the third and fourth centuries found there suggest that the miners were tied to their working-places, and were slaves or convicts. At the Parys Mine in Anglesey, cakes of copper associated with native villages imply that the ore was gathered by natives and smelted in villages for collection at a central depot. This island, after the expulsion of the Druids, was treated as a temple estate with a fixed tribute. The fort at Segontium (Caernarvon) and the provision of a naval station Caergybi at Holyhead harbour as a protection against pirates and Irish slave-raiders illuminates the importance the Romans attached to this rich, large copper deposit.[17]

It is not clear when the Romans penetrated Cornwall. During Caesar's conquest of Gaul anti-Roman refugees crowded into south-west Britain and this cannot have encouraged Roman adventurers. When the tin mines of Spain and Portugal were exploited, the demand for Cornish metal almost ceased, but when Iberian supplies began to diminish, interest in Cornish tin revived, so that the tin trade became prosperous again in the middle of the third century. Up to then, though the Romans took tribute, official control does not seem to have extended to the tin industry. However, the remains of a Roman villa were found at Illogan, north of Camborne. In the third century, there was a new development in the wide use of pewter table services. Officially-stamped blocks of pewter have been recovered from the Thames at Battersea.[18]

Some gold was washed from Welsh rivers and from rivers in Dumfriesshire and Lanarkshire but the amount was not significant. The only Roman gold mine of importance was at Dolau Cothi, a remote point in the province near Pumpsaint in Carmarthenshire. Outcrops of gold-bearing quartz were attacked and two adits led to underground workings in the hillside. A seven mile leat, a manmade ditch, brought water from the river to the mine for dressing the ore.

When the Romans left the Britons to look after themselves the island was raided and later settled by Anglo-Saxons and Danes, and mining was neglected. Cornwall, however, continued to produce tin; a ship carrying corn to Britain to relieve a famine returned to St John the Almoner, Patriach of Alexandria, with a shipload of tin in A.D. 618.[19] Also many Anglo-Saxon bracelets and ornaments have been recovered from old alluvial tin workings. Tin works were probably active both before and after Athelstan's conquest of Cornwall in A.D. 937, though constant Danish raids must have made mining precarious.

Before the Norman Conquest little is known of other metal mining, though old records show that the Saxons mined and smelted lead at Wirksworth and Castleford. The Abbey of Repton owned some of the Wirksworth mines in A.D. 668, possibly because Repton was the royal centre of Mercia.[20] When it was destroyed by the Danes in A.D. 874 the mines came under a Danish earl. The Odin mine at Castleford supposedly takes its name from Danish workers.

Doomsday Book records three lead mines at Wirksworth and one each at Bakewell, Ashford, Matlock and Crich. Iron mines are also mentioned, but not tin. Presumably this is because tin was considered to be royal property. After the Conquest more attention was paid to mining. Tin, lead and wool were the principal exports of the Middle Ages and internationally were scarcely less important than the spices of the East.

Though woefully incomplete, the best documented mining field of this period is Devon and Cornwall. Up to the end of the twelfth century the smelters were self-employed craftsmen, but by 1586 capitalists employed a number of blowers. By the second half of the fourteenth century the tin miners were constituted under their own peculiar laws and customs, held their own courts and paid taxes direct to the King. John in 1204 gave them their first charter. Edward the First's charter of 1305 separated the tinners from the rest of the community and the stanneries of Devon were parted from those of Cornwall. The Cornish, speaking a Celtic language and with different traditions and manners, were thus separated from the men of Devon, who probably spoke the English of Chaucer. The previous banishment of the Jews in 1290, however, made the promotion of mines difficult. Edward III made over the Duchy to the Prince of Wales and his heirs and so it remains.

The tinners were a thorn in the side of the local lords, and complaints against them were many. In 1361 John de Treevres complained to the Prince that they had destroyed his crops and sixty miners had dug for tin and diverted streams so that nothing was left but stones and gravel, and he prayed that 'for the love of Christ you may be pleased to ordain a remedy'.[21]

Under the Charters, employers, shareholders, dealers and all artisans claimed the privileges of the Stannary Courts. Commissioners were frequently sent to ferret out the wealthy tax-dodgers who fought vigorously to retain their privileges for three hundred years. Henry VII with his cupidity declared the Charters forfeit and only relented in 1507 on payment of a fine of £1,100. From 1751 masters, managers and agents of smelting houses paid heavily for breaches of the regulations, but as late as 1855 anyone connected with the tin industry could only be sued in the Stannary Courts.[22] The 'free tinners' were a robust, independent and at times a turbulent body under the special protection of the Crown, but they suffered badly at the hands of the middlemen and smelters.[23]

The Black Death of 1348 caused metal production to diminish all over the country, but a couple of centuries later there began a great revival of interest in the metal industries of Great Britain. This was in part due to the influence of Saxon miners on English practice, which increased in late Tudor times, but it was more prolonged and deeper than is generally recognized. Edward I had originally introduced them as early as 1299 and his son Edward II brought Herman, a German, to work at Dulverton. By 1359 Germans were working the lead mines in Alston Moor. In 1452 Henry VI brought skilled miners of Saxon origin from Bohemia and Hungary and Edward VI granted mine leases to Germans in Northumberland and Westmorland. In Elizabeth's reign William Humphrey, Assay Master of the Mint, 'by his great endeavour, labour and charges has brought into her realm of England one Christopher Shutz, an Almaigne, to dig and mine for calamine stone (carbonate of zinc)'. He was given a patent for making 'latten' or brass in 1565. Lord Burleigh fostered metal mining to reduce imports and boost exports against the threats of Spain. Queen Elizabeth in 1562 sent to Germany for miners for sorting, sieving and washing copper ore at a mine near Kendal. She granted all mines and minerals in her dominions to two corporations, the Mines Royal and the Battery Works, for all time. They were terminated during the Commonwealth but reinstated after the Restoration. She also brought German miners to Derbyshire to introduce better methods. They were granted concessions and acquired property but have left little trace. In 1602 Sir Francis Godolphin, on the advice of a German ore-dresser, profitably redressed tailings left by the Cornish miners.[24] In 1638 German miners introduced gunpowder at Ecton copper mine and by 1733 gunpowder was in use in Cornish mines. Many Saxon words are still used in British mining, such as stope, stull, sump, stamp and trommel, and Cornish surnames such as Doidge, Knuckey and Knapp call to mind the men invited into England to impart their special skill to the metal mining industry of England over the middle centuries.

In Charles II's reign, the London Lead, or 'Quaker' Company was incorporated to mine lead and silver. They were paternalistic and centuries ahead of their time concerning the welfare of their employees.[25]

In Derbyshire lead mining was the major industry from Roman times, and its curious laws and customs still control it. The ancient Barmote Courts governing the working of lead in that county can be traced back to 1289 with an origin even more remote, as the word itself is of Saxon origin.* Edward Manlove in 1653 published his 'Liberties and Customs of the Lead Mines within the Wapentake of Wirksworth in the county of Derby composed in meter' so that the illiterate miners could memorize it. The Great Barmote Court was held twice a year with a jury of twenty-four men with a wide knowledge of mining customs. All judicial matters had to be brought before the Barmote Court and the coroner had no jurisdiction over a miner's corpse.[26]

The Derbyshire miners always held strictly to their ancient priviliges, but these were not codified until as late as 1851-52. They confer power on any of the Queen's subjects to search for lead in the High or Low Peak in any place, not a garden, orchard, churchyard or highway, regardless of the owner's wishes. The finder can register his claim to be allotted land for shafts, dressing plant buildings, dumping of spoil and passage to the nearest highway. The last leads to the curious and still existing custom of the Barmaster and two of the body of the mine walking side by side with arms outstretched from the mine to the nearest public road, marking the edge of the approach with wooden pegs. The miners were obliged to measure the dressed ore for sale in a dish, nearly always of wood, checked twice a year against a standard brass dish of 472 cubic inches capacity kept at the Moot Hall at Wirksworth and dated 4 October 1513. Tithes varied from one-tenth to one-fortieth and were made by setting a dish of dressed ore aside.[27]

* Saxon, Bergmote = Mountain Court.

During the seventeenth and eighteenth centuries prodigious efforts were made to drain the mines by soughs, almost horizontal self-draining galleries set at a point as low as possible in the valley. Cromford Sough was two miles long and cost £30,000, and Hillcarr Sough from the river Derwent at Darley Dale to drain the Alport and Youlgrave mines was four miles long and cost £50,000. Begun in 1754, it took twenty-one years to construct.[28]

Daniel Defoe (1668-1731), author of *Robinson Crusoe*, walking through Derbyshire, noticed a miner climbing out of a lead mine, pale, thin and as grey as the ore; yet another miner he met lived with a wife and five bonny children in a hillside cave which he described as a neat, clean and comfortable home.

Tin lode-mining in Cornwall did not start until medieval times, when the output from the great eluvial deposits began to decline. It probably started on the exposed tin veins in the cliffs of north Cornwall, a good example of which is Cligga Head near Perranporth.

The first great upsurge in metal mining was in Cornish copper in the eighteenth century when, from 1730 to 1865, Cornwall produced nearly eight million tons of copper ore worth over £50 m.[29] At the end of the eighteenth century there were more than seventy mines producing copper ore. Of these Dolcoath was the most important; it produced only copper down to a depth of 1,200 feet but, after passing through a zone poor in both copper and tin, became a tin mine down to 3,000 feet. It was the chief underground tin mine in Britain and, in its time, in the world.[30]

Copper production in the United Kingdom after Roman times dates back as far as about 1250 but it was not of importance until copper smelting began at Landore in the Lower Swansea Valley about 1717, using locally mined coal. There were thirteen smelting works there and by 1850 the sulphurous fumes had killed all trees, grass and heather. In its prime Swansea smelted 90 per cent of British copper ore, and Britain produced three-quarters of the world's copper metal from ore mined in Cornwall, Devon, Anglesey and Ireland.[31] It remained the principal copper producer until 1851 when a steady decline began, caused by the discovery of huge, rich copper deposits in North America. Two decades later the industry died, although the last copper mine at Glasdir, Merionethshire, did not shut down until 1914. Swansea smelters then switched to zinc smelting, most of the zinc ore coming from lead mines. Nothing remained in the valley where life had been killed but 1,200 acres of bare unsightly dumps showing the desolation that the metal industry can leave behind, but it is now rehabilitated in part.[32]

The tin boom was later than that of copper for in 1801 there were only 75 tin mines in Cornwall, yet by 1862, according to Williams's Mining Directory, there were 340 working tin mines employing 50,000 men.

Although it is now forgotten, the middle years of the nineteenth century were the heyday of the British metal mining industry. It produced three-quarters of the world's copper and half the world's lead as well as 60 per cent of the world's tin. During the second half of the nineteenth century domestic metal mining was rapidly declining, and mine-owners closed the mines when the internationally controlled metal prices were low and opened them again when a rise in price made it possible to work the mine profitably. This stop-and-go policy bore hardly on the working miner, but for generations in most parts of the country the miner was also a smallholder. So when the mines closed he took to farming: a precarious existence. In North Wales during this period the wages on the dressing floor were three shillings a day for a man and one shilling for a boy. Also at this time there was a general exodus of miners from Cornwall, mainly to the United States of America when the 1849 gold rush made that country so inviting. Today, according to A. L. Rowse, himself a Cornishman, there may well be eight times as many people of Cornish descent in the United States as there are in Cornwall.[33]

At the end of World War I an excellent report to the Government was produced by a committee under the chairmanship of Sir Lionel Phillips, a South African mining magnate, followed in 1920 by another report by a committee under the chairmanship of H. B. Betterton M.P. on the non-ferrous mining industry.[34] Both gave past outputs in great detail, showing the decline during the second half of the nineteenth century. The latter report (Para 120) stated that the industry was unlikely to revive without direct state aid. Up to and after World War II committees met on the subject, but no material benefit was accomplished, more especially as regards relief from taxation.

Between the wars there were a number of attempts by important mining houses to create or reopen mines in England and Wales. Two lead mines, Halkyn in Flintshire and Mill Close in Derbyshire, together produced 68,000 tons of high grade concentrates in 1934. Unfortunately metal prices were low during their years of activity, and both closed down, not for lack of ore but because of incoming water, the common curse of mining in limestone country everywhere.

Under the threat of World War II there was a frantic search for useful sources of base metals in the dumps of old mining areas of England and Wales, Anglesey and the Isle of Man. Only one such dump was processed, at Nenthead in County Durham for zinc.[35] During the war a committee under the chairmanship of Sir William Larke sat to study the possibility of domestic sources of non-ferrous metals. But new mines are not made in a day; they take years, sometimes many years, before becoming productive.

Since World War II efforts in both England and Wales to create or reopen base metal mines have been made by mining houses with ample financial resources, but all but one have so far failed, and today there are no active copper, zinc or lead mines, though a small quantity of lead ore is got as a byproduct from mining fluorspar. There are today only three working tin mines, South Crofty, Geevor and the recently opened Wheal Jane, near Truro. The marvel is that Cornwall is still producing tin after four thousand years.

It must be stressed that mining and farming are the two primary industries where man competes not only with other men but with Mother Nature. On the well-cultivated fields of Britain crops have been renewed and beasts reproduced for many centuries. Every mineral deposit, however, is a wasting asset, its sole capital is the mineral itself and as it is consumed in mining its life is successively shortened until the supply is ended; the mine is then dead, and its specialized machinery of little or no value.[36] Nevertheless no British government has given the industry special financial treatment as has our neighbour the Republic of Ireland, copying the mining laws of Canada and other countries where metal mining is a major industry.

Again, there are in this country two highly subsidized branches of primary industry. There are hundreds of thousands of coalminers and farm workers with votes but only hundreds engaged in metal mining. So any government, right, left or centre, can afford to ignore metal miners and has done so. Nevertheless, the importation of tin, lead, zinc, copper, nickel and tungsten cost us nearly £600 million in 1968.

This country is almost unique in that mining rights are in private hands. As far back as 1944–45 the late Professor W. R. Jones eloquently pleaded for the nationalization of mineral rights in Great Britain.[37] His was a voice crying in the wilderness! Recently the national press had loudly ventilated its appeals for more generous treatment for those international mining houses trying again to create viable mines in Great Britain. Headlines are various: 'Archaic Tax Law Slows Down Mining Boom', 'Government May Ease Mineral Search Problems', 'Tory Threat to Kill Britain's Mining Industry', 'Landowners Hold Up Mineral Mining', 'U.K. Mineral Search is On', 'Fair Price for Mineral Rights'; and many more.

Currently there are important prospects afoot concerning lode gold in Merionethshire; copper in the Hermon valley; copper, lead and zink in North Wales; tin in Cornwall at Mount Wellington, in Mounts Bay and at Geevor in reopening the old Levant Mine under the sea. The present government's policy of reducing public spending by the abolition of the 40 per cent investment grant is counterbalanced by grants up to 35 per cent of a company's exploration and feasibility costs.

The industrial archaeology of the industry is enormous. There are innumerable ruins of old buildings, entrances to old mines, the Racks of Derbyshire, thousands of old mine dumps and many old smelting sites. They range in age from the Celtic Chun Castle in Cornwall and the Roman gold mine at Dolau Cothi to the foundations of surface plants at Wanlockhead in Dumfriesshire and Florence Mine in Cumberland, both closed in recent years. Geographically they spread over the whole country. The tall gaunt ruins of engine houses of old copper and tin mines from Chacewater to Camborne in Cornwall, the ruined lead mining and smelting sites of the Yorkshire dales, County Durham from Teesdale to Blanchland, Alston Moor, Cumberland and Westmorland, north, south, west and east Wales to Minsterley in Shropshire with its lead mine going back to Roman times; copper mines at Alderley Edge and at Parys in Anglesey, the old smelter sites near Swansea and the gold mine of West Wales: all these are among the innumerable ruins and relics of a famous past.

When the Romans ruled, metal mining was our major industry, and just over a century ago we were the greatest provider of base metals in the world. The proof is there to see from Lanarkshire to Cornwall and Anglesey to Kent.

CHAPTER TWO

Gold

Britain was never so important a gold producing country as prehistoric Ireland. However, Britain did produce some gold in ancient times, as both Strabo the geographer and Tacitus the historian mention gold and silver as products of the province of Britannia. In Britain, mines producing gold and silver are called Mines Royal.[1] The earliest record of a sovereign enforcing this right was Henry III in 1262 concerning a gold-bearing copper mine in Devonshire. Henry VIII in 1528 appointed Joachim Hoegstre as principal surveyor of all mines in England. He was unwelcomed, his house was attacked, but finally he and his family settled peaceably in this country. He was the father of Queen Elizabeth's Daniel Houghstetter. Queen Elizabeth on the advice of her council, sent for experienced German miners and on the 10 October 1562 granted the mining in eight counties to Daniel Houghstetter, Ludwig Haus and Hans Louver, who brought with them several mining experts from Gastein in the Tyrol. Among other places in Cumberland and Westmorland, they visited copper mines near Keswick. By 1566 copper was produced there.

The Society of the Mineral and Battery Works was founded by Royal Charter in 1564. Shortly afterwards a second Corporation of the Mines Royal was established for the management of the royal mines in which the German Daniel Houghstetter took a leading part. The two corporations monopolized English mining as regards gold, silver, copper and calamine and manufactured brass and iron wire.[2] Copper was mined on a considerable scale near Keswick and a thousand men were employed there as late as the Civil War.

In 1668 the two corporations were combined as the Society of the Mines Royal. They did not operate mines on their own account but granted leases to capitalists to dig for ore in various places. Among

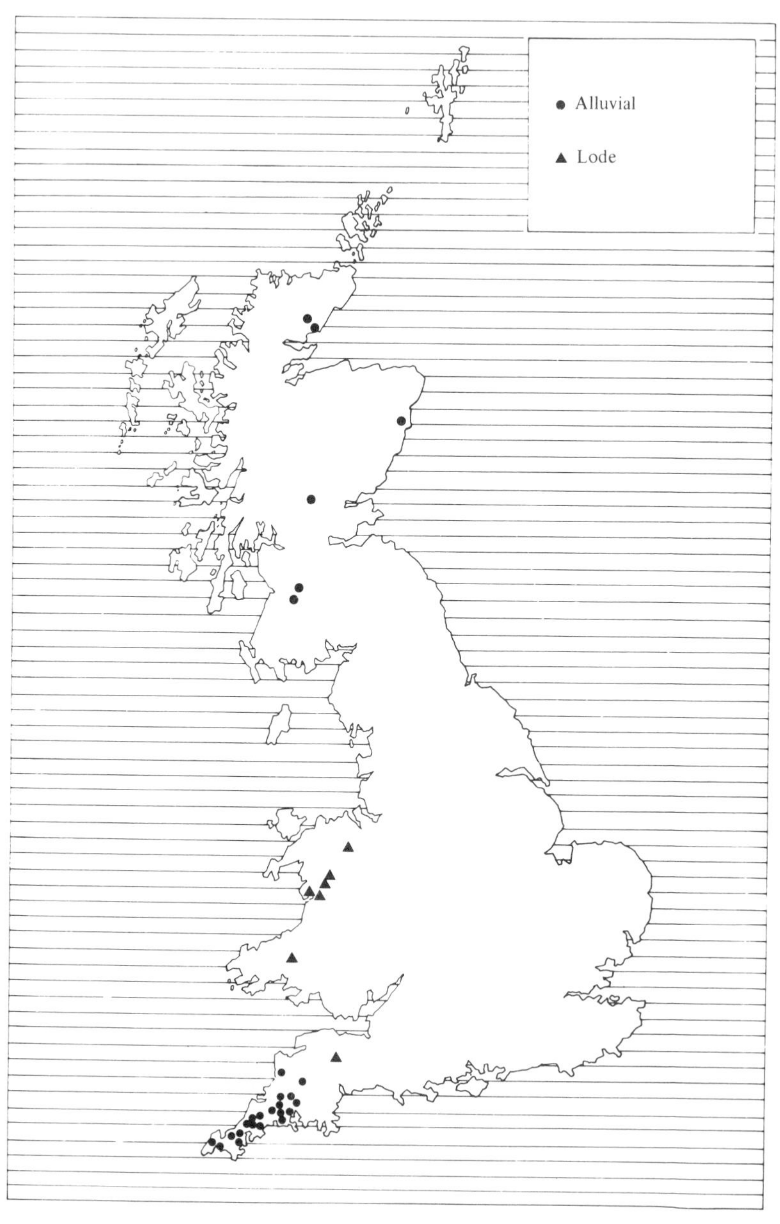

Figure 2 Gold deposits

their lessees were such notables as Hugh Myddelton, Thomas Bushell, Sir Carbery Price and Sir Humphrey Mackworth. During the seventeenth century its influence was pernicious, tending to check all private mining enterprise and causing landowners to conceal rather than exploit mineral deposits.[3]

Besides establishing the Mines Royal Queen Elizabeth I also in 1568 sued the Earl of Northumberland and won a famous lawsuit which resulted in the resumption to the Crown of the rich copper mine of Goldscope (Gowd-Scalp) at Newlands, Keswick, where gold had been found in the reign of Henry III. Actually the mine was resumed on account of the high silver content rather than the gold, for a writer in 1709 remarked: 'The ore got by gin under level was so rich in silver that Queen Elizabeth sued for it and recovered it from the Earl of Percy for a royal vein.'[4]

The privilege of working a Mine Royal had to be sought at Court, where such diverse characters as Prince Rupert (1619-82) and two royal portrait painters, Nicholas Hilliard and Cornelius de Vos, were involved.[5] The Company's activities were suspended during the Commonwealth but revived after the Restoration. Formerly there was considerable dispute as to what constituted a royal mine. Some authorities considered it to be a principle of common law that if *any* gold or silver was found in metal of a baser nature that was sufficient. Others considered that a mine was not to be deemed royal unless the quantity of gold or silver exceeded in value the other metals mixed with it. In 1568 Edmund Plowden (1518-85) in his judgement contended that if the royal metals bear the expense of extraction the whole should belong to the Crown, if otherwise to the owners of the base metals.[6]

The right of entry in search of royal mines became oppressive in the extreme, and no damages were paid. There was universal mistrust and a remedy was called for. It was eventually obtained, for in 1689 an Act was passed declaring that no mine of tin, copper, iron or lead should be taken to be a royal mine although gold or silver might be extracted from it.

This provision was considered insufficient, and another act was passed in 1693-94 (the fifth year of William and Mary repealing a statute of the fifth year of Henry IV). In this it was enacted that all owners or proprietors of mines in England and Wales in which there was copper, tin, iron or lead should hold and enjoy the same mines and ore, dig and work the same, notwithstanding that such mines or ore should be claimed to be royal mines. However, the Crown retained the right to purchase the ore of any such mines. The Crown still has the right to all gold or silver mined, hence a percentage is paid to the Crown of gold or silver taken from such mines, or a tax in lieu – a royalty.

During the gold mining boom in Merionethshire, venturers were hindered by claims of the Mines Royal Corporation. Finally, in 1860, it was settled by a ten per cent royalty for mines on Crown land and a five per cent royalty for those on private land. A Customs and Excise report in 1930 stated that standard rates had been fixed in 1896 entitling the Crown to all gold and silver from mines whether owned by the Crown or privately. Crown land is usually leased for gold mining at a fixed rent and a royalty of four per cent of the gold obtained. Private owners accept an annual rent and a two per cent royalty to the Crown.[7] Application for licences to prospect for, or to work, a royal mine must be made to the Crown Mineral Agents and the Crown Estate Commissioners. The Crown holds the gold mineral rights over the whole of Britain except Sutherland.

In Cornwall the numerous writs and grants from the reign of Henry III to Elizabeth I to various grantees indicate that it was thought profitable to work for gold there, but not a single ounce of gold was recorded in that period. J. H. Collins, the great authority on West of England mining, lists forty-two Cornish parishes where gold has been found, from Callington near Calstock on the River Tamar to St Just near Land's End. He gives thirty examples of native gold found in Devon and Cornwall. The largest was a smooth nugget found in the Carnon stream at the head of the Restronguet creek five miles north of Falmouth. Weighing 1 oz 16 dwt 8.6 grains, it is now in the Royal Institution of Cornwall Museum in Truro.[8]

Beare's *Bayliff of Blackmore*, written about 1586, records that:

> About thirty-six years ago my fortune was to be present at a wash of a tynne work in Castle Park by Lostwithiall, where there was a certain gentleman present, whom I could name, gatheringe out from the heap of tynne certain glorious cornes affirmed them to be pure gold which the tynners permitted him very gently as they will gentilly suffer any man to do most chiefly if any liberalite will be shown amongst them to the value of one 2d to drink then shall you have them very diligently go to their buddle themselves and seek out amongst their cornes of tynne which they call 'rux' the finest and more radient cornes and present them to you.[9]

Richard Carew's *Survey of Cornwall* in 1602 remarks: 'Tynners doe also find little hoppes of gold amongst their owre which they keep in quils and sell to the goldsmithes oftentimes with little better gaine than glaucus exchange.' (During the siege of Troy a Lycian prince exchanged his gold armour for the brass armour of Diomedes.)[10] To preserve these "prills" the tinners carried a quill with one end fitted with a wooden plug into which small receptacle they carefully dropped the tiny bits of precious metal. It was a tinner's perquisite and occurred frequently enough for the men to be paid a lower wage than when working where there was no gold.[11]

Gold from Ladock assayed gold 92.34 per cent, silver 6.06 and gold from St Austell's Moor assayed gold 90.12 per cent and silver 9.05 per cent.[12] It is said that all the gold from Cornish tin stream works only amounted to a few pounds, but Collins reckons that over the centuries three-quarters of a million tons of eluvial black tin was produced, and if it only carried one pennyweight of gold to the ton it gave a not inconsiderable 37.500 ounces of gold.[13] How much of this was recoverable and how much recovered will never be known.

In Devon gold was found in the eluvial deposits of streams flowing from Dartmoor where the existence of gold had been assumed for centuries. At the beginning of the nineteenth century a miner named Wellington found gold at Sheepstor in South Dartmoor. From time to time he brought small quantities to a silversmith in Plymouth, but only to the total value of about £40.

In 1797 the Rev. R. Polwhele, author of *History of Devonshire*, wrote: 'From the copper of a mine in the parish of North Molton some gold has been lately extracted.' In 1850 speculators were attracted to the district and obtained a lease from Lord Poltimore to extract gold from the Poltimore mine at Heasley Mill.

North Molton lies on the southern slopes of Exmoor and in 1852 the gozzan* ores of both the Britannia and Poltimore mines were found to contain payable gold. The discovery, so soon after the gold booms of California in 1849 and Australia in 1851, attracted extraordinary attention. The Editor of the *North Devon Journal* in June 1854 stated: 'A rich harvest for shareholders . . . a hundred tons of ore is at grass which contains 27 ounces of gold per ton, each ounce worth nearly £4.' Actually the first trial yielded 26½ ounces of gold from 20 tons of ore, but the

*Decomposed rock indicating an underlying metallic vein.

subsequent trials of 50 and 75 tons yielded an average of only 16 dwt of gold. These gozzans were from four to ten feet wide, dipping almost vertically, and there is evidence that they were worked for copper in remote times. The ore was highly mineralized friable ironstone with a copper content. On the west side of the river Mole it was brown and contained 8 dwt of gold but on the east side it was reddish and contained 17 dwt to the ton of ore.

The Britannia mine is three-quarters of a mile north of Poltimore and grains of gold had been found there prior to 1822 by a Mr Flexman of South Molton. The gozzan there was more siliceous, but both are derived from the decomposition of gold-bearing iron pyrites. The total value of gold produced to 2 November 1853 was only £581 5*s* 1*d*. The excitement was shortlived and in 1856 the lease was purchased on behalf of Brocklebanks, the Liverpool shipowners, and all the plant and machinery was sold for £2,625. Thus the North Molton gold rush became history. In the Natural History Museum, South Kensington, there is a specimen from these mines with small particles of gold clearly visible in the brown siliceous ironstone.

The most historically interesting gold mine in Britain was the mine of Ogofau (caves) or Dolau Cothi (loops of the river Cothi) half a mile from the village of Pumpsaint (five saints*) in Carmarthenshire (see Plate 2). Here the Romans worked a deposit of free gold in quartz veins and the oxidized portion of goldbearing pyrites. They constructed a manmade watercourse seven miles long between the 600 and 800 foot contours, mapped by a survey in 1840. Much of it was cut out of the solid rock to provide water drawn from higher up the river to operate primitive stamps to crush the ore and for dressing the crushed ore. This leat fed a still recognizable reservoir 100 ft by 30 ft cut out of the rock on the hillside above the main working.

The Romans exploited the deposit by both opencast and underground workings. Besides the main opencast there was another at Pen-Lan-Wen 800 feet to the west, and a third, Cwm Henog, 1300 feet to the east. The ancient main adit† is typically Roman with sides tapering towards the base, shaped for labourers to carry the broken ore out of the mine in skin sacks or baskets on their shoulders. This gallery was cut by hand tools with admirable precision.[14]

* A local legend is that five saints born at one birth rested their heads in the hollows of the stone mentioned below.

† Self-draining gallery.

A mining engineer has estimated that the Romans excavated four million tons of rock. In the main opencast there were six bands of ore striking across it sufficiently wide and rich for them to work it in one excavation rather than mine each vein separately. These ancient workings were excavated down to the level of the river Cothi, probably all in oxidized ore. The Romans mined no deeper because of the difficulty of recovering gold from the sulphide zone and the trouble of coping with incoming water. In recent times the old Roman underground workings were broken into and the old stope faces were found to be lined with the embers of wood fires lit to heat and fracture the quartz, proving that fire-setting was practised.

The value of the gold won in Roman times will never be known, but the Emperor Tacitus (not to be confused with the historian) in A.D. 275 claimed that Welsh gold was among the satisfactory rewards of victory. Historians think that because of relics found there, the remoteness of the mine and labour supply difficulties, that public baths and other amenities were provided to retain a contented labour force as they undoubtedly were at a similarly remote Roman copper mine at Ajustral in south-west Portugal.

Among the relics found was a wooden water wheel now preserved in Cardiff Museum, but an excellent wooden ladder was too water sodden to be salved.[15] Other Roman finds include a wooden panning dish found at some distance underground. There is a block of coarse sandstone on the green in Pumpsaint village showing cavities on each of the four sides. A two to three feet long depression, eight to ten inches wide, and four to six inches deep, created elliptical grooves. The stone was almost certainly used as a mortar in which the quartz was crushed or ground. It would appear that the grooving was made by a suspended pestle with a reciprocating motion like the old Australian dolly. Archaeologists have found necklaces, wheel-shaped clasps and bracelets in the area, all of gold.

The village of Caio (*caius*=a camp) a mile south-west of Ogofau is surmised to be the site of the mine labourers' settlement in Roman times. There is no evidence of mining here between Roman and recent times. William Camden in his *Britannica*, first published by Newbury in London in 1586, mentions the Ogofau caves as a curiosity and speculates on their origin. In 1839, Sir Roderick Murchison visited the site and realized that they were Roman workings but did not know the metal mined. In 1846 Warrington Smyth, later to be the first Professor of

Mining and Mineralogy at the Royal School of Mines rediscovered gold in the workings, but only a small amount of fossicking was done. From 1889 to 1891 the South Wales Goldmining Company worked there but with poor results as the yield was only 4 oz 19 dwt of gold.

In 1903 a Cornishman, James Mitchell, made the first serious attempt to resuscitate the mine. He installed a fivestamp mill, restricting his work to scratching round the old surface exposures. In 1909 he sold his rights to Cothi Mines Ltd and stayed on as manager. By 1910 he had sunk a 100 foot shaft and carried out a little development work,[16] but broke into ancient workings, with the result that the mine was flooded and the company became insolvent. Mitchell's report of 4 November 1909 showed the shaft to be sunk down to 96 feet, with 4 feet of quartz giving 11 dwt to the ton. Like the Romans, Mitchell stuck to the lode when and where he found it, but worked upwards, whereas the Romans went down on the lode. Samples of ore from the mine were exhibited at the Imperial International Exhibition at Shepherd's Bush. The Company was liquidated on 22 May 1912. Thus ended the work of James Mitchell.

The mine was abandoned for the next two decades, when it was acquired by a syndicate in December 1934, and taken over by Roman Deep Ltd. Meanwhile from 1909 onwards it had been offered a number of times by several parties, brokers and mining experts, to one of the largest international gold mining groups. The ore was examined thoroughly by their metallurgical subsidiary to find that the presence of galena (sulphide of lead) and arsenical pyrites made the treatment complicated, necessitating fine-grinding, flotation and roasting to produce a concentrate containing 95 per cent of the gold content. It was offered again as late as 1939, but was turned down as too highly speculative and over-capitalized.

Meantime in December 1934 Roman Deep Holding Ltd acquired the mine from Roman Deep Ltd for £300,875 in fully paid shares, acquiring the mining rights over 3,500 acres. Roman Deep Holding Ltd registered a subsidiary company, British Goldfield (No. 1) Ltd in April 1937, with rights over 230 acres to work Ogofau mine. Mitchell's shaft was deepened to 480 feet and shaft stations cut. The treatment of the mine labour, mostly local farm hands, was good and the camp provided for them developed into a board-residence establishment with a steward, his wife as cook and a club licence.[17] A Welsh writer considered it to be immensely superior to the barracks for miners in other Welsh mining concerns.[18]

Milling the ore started in January 1938. The reasons for failure appear to be that the original mill flow sheet was only for free gold. No provision was made for the treatment of sulphides. Nevertheless when a flotation plant was added later towards the end of the enterprise, recoveries were made of over 98 per cent of the contained gold. Failure was also attributable to several other causes. The mill feed was disappointingly low in values because of dilution by country rock as a result of mining irregular stringers of ore instead of sticking to the lode as the Romans and Mitchell had done; production of a saleable product was started before an economic tonnage was fully blocked out; the pickings of the ancient tailings dumps were ignored; the lease forbade any building so placed as to mar the scenery, thus causing the mill and powerhouse to be sited 250 feet above the shaft collar instead of below it near the main road; power from the national grid was not available; the concentrates were sold to Seattle, U.S.A., thereby incurring high handling, freight and insurance charges. But the main cause was starting the enterprise at a time when many profitable overseas mines were being compulsorily closed by the threat of war.

Underground drilling and crosscutting ceased by the end of 1939. Early in 1940 a rough geological surface survey was made of the whole area before the property was abandoned, the lease returned to the landlord and everything sold lock, stock and barrel. So the mine, worked successfully eighteen centuries ago, died, and the only evidence today is a concrete slab sealing the shaft collar. The Romans had the best of it at Ogofau.

According to the *Observer* of 20 February 1972, Dolau Cothi, owned by the National Trust, is to become a centre of archaeological research and Rio Tinto Zinc is helping to finance the project. It was the most technologically advanced Roman mine in Britain and it is guessed that it produced a hundredweight of gold a week which, at today's value, is about £30,000 worth. The greater part of this gold was sent to the Imperial Mints at Lyons and Rome. The tanks for dressing the ore mentioned above had walls of solid rock 55 feet thick and held a quarter of a million gallons of water. The Romans exploited this deposit immediately after A.D.75 and it is surmised that they imported experienced miners from the Asturias in north-western Spain.

There is an old copper mining district in Merionethshire occupying about twenty-five square miles north of a line from Dolgellau to Barmouth and to the north bounded by the somewhat sinuous contact of the Ordovician and Cambrian rocks. North of a mile below the Rhaiadr Mawddach waterfall no gold discoveries of any importance were made.

The existence of gold in the district was known in the reign of William IV but was not commercially exploited until 1843.[19] A Mr Robert Roberts claimed to have had samples assayed in 1836 that proved to be rich in gold and silver, but he was not aware that such ore could be worked in North Wales, so made no further effort until 1843. Several years after 1836 he told Arthur Dean, who was called in by James Harvey, the owner of the Cwm Eisen mine, to make an expert report. All three claimed to be the discoverer of gold in North Wales but Dean read a paper before the British Association in 1844 containing the first public notice of the discovery.

An attempt to raise capital to work the gold mines in 1846 was ridiculed. Early in 1847, however, the North Wales Silver Lead Copper and Goldmining Company was floated, with 12,500 shares of £10 each, to work the veins at Vigra, Clogau, Tyddyn-Gwladys and Dol-Frwynog. The first two were then being worked for copper.

Before January 1849 the first extensive trials for gold were made at Cwm Eisen and 7 pounds of gold were obtained from $10\frac{3}{4}$ tons of concentrates produced from 300 tons of ore. On 16 August 1853 gold was discovered at the Prince of Wales mine and the same week detected in an old dump at Vigra mine.

In 1853 an impetus was given by the introduction of the Berdan machine for gold recovery, and in 1854 a single rock crushed at Clogau produced £25 worth of gold. In that year there was a gold rush that kept the county in a state of excitement for two years. The mines worked at this time were in the vicinity of the Rhaider Mawddach waterfall. The boom died rapidly.

Eight years later a vein of gold was discovered at Clogau, but after the sad experience of 1854-55 the people concerned were, if not more honest, at least more prudent, and employed 'an excellent miner', one John Parry, to conduct the development work—a new departure in a district where copper lodes had been worked, according to Warrington Smyth, in an astonishingly barbarous manner.[20]

Under Parry the ore was won by four men on development work with another four stopping. For extracting the gold there was one Britten machine and one American Berdan machine, a leftover from the 1854 gold rush. Both machines combined grinding and amalgamation. Only hand picked ore was fed to the Britten machine. The results for fifteen months from 11 January 1861 to 4 April 1862 were:

	Ore treated		*Gold recovered*	
	tons	cwt	oz	dwt
Britten's Machine	9	14	3,624	18
Berdan's Machine	650	8	686	9
Total	**660**	**2**	**4,311**	**7**

These results proved the outcrop to be rich, and justified further development, but after the rich outcrop was worked out the mine changed hands and suffered many vicissitudes.[21]

The success at the Vigra and Clogau lodes gave rise to a vigorous prospecting programme. The best result was at Gwynfynydd (white mountain) mine which claimed to be the most successful enterprise in the Mawddach valley. Cwm Eisen and Dol-Frwynog, the earliest producers, were never remunerative. During 1865 many thousands of pounds were spent on useless ore-dressing machinery. In that year, however, the Clogau mine paid £22,575 in dividends and had produced gold valued at £43,783 in three years.

After 1866 gold mining languished for nearly two decades. Small patches of rich ore were found, worked out and mining suspended. In 1870 the total production was 191 oz of gold of which 165 came from Gwynfynydd. During the year following not an ounce of gold was produced.

In 1888 a rich shoot was found at Gwynfynydd, the Morgan Company floated and over £35,000 worth of gold was produced in two years, but the shoot was soon exhausted and operations suspended.

The St David's Gold and Copper Mines Ltd started development operations in a systematic manner in 1898. It controlled the two most important mines, Clogau and Gwynfynydd. In 1899, from 1,200 tons of ore crushed it produced 2,831½ oz gold; in 1900, from 15,833 tons of ore, 11,284 oz gold; in 1901 from 15,500 tons of ore, 5,537 oz gold; from January to June 1904, 6,762 tons of ore produced 7,797 oz gold. In 1900 the net profit was £39,729 and a dividend of 60 per cent was paid. Royalties paid to the Crown were £2,038. 7*s* 7*d*, or 2*s* 1*d* a ton of ore crushed.

At St David's the gold bearing veins varied greatly in width from nine feet to a mere streak, sometimes showing visible gold in dark bluish quartz with inclusions of country rock. The white quartz was barren. All the ore was treated as free-milling because the sulphides contained no gold. All the gold veins in the region are extremely patchy rich shoots of ore separated by runs, sometimes extensive, of barren quartz. Normally the gold contains about 9 per cent of silver. Two samples from Clogau assayed 90.16 per cent gold, 9.26 per cent silver and 89.83 per cent gold and 9.24 per cent silver.

St David's Mine was attacked from three adits, one a crosscut which intersected the lode 500 feet below surface; in this gallery was the tramway to the head of the ropeway, which had a capacity of 24 tons an hour. When in mid-1904 the south lode was opened up the mine proved to be rich, the vein from 10 to 12 feet wide, 2 feet of which was the bluish-grey quartz typical of the Clogau lode and in which most gold was found. A seven-ounce nugget was found in this lode. The recovery of copper concentrates as a source of additional revenue in an Elmore plant was not successful as the copper values were too low. The tailings from the battery plates gave only 0.2 per cent copper.

In Merioneth there were twenty-four recorded gold mines to the west and north of the Mawddach valley, a country of knife-edge hills, beautiful ravines and quiet valleys. In some years 10,000 ounces of gold were won, and in others less than ten. The total output from the discovery in 1843 to 1904 was 88,882 ounces of gold.[22]

There was alluvial gold as well as lode gold. As early as 1852 Frederick Walpole and Sir Augustus Webster panned a considerable quantity from the Mawddach river. In the early 1870s Californian and Australian miners were attracted, and, owing to the extremely low water, worked the river bed profitably. Since then many amateurs have washed a little gold below Pistyll-y-Cain and in quiet pools below the waterfall. In 1970 Rio Tinto Zinc made a seismic survey of ten miles of the Mawddach Estuary in a study for dredging alluvial gold but subsequently abandoned the project.

All the gold was melted and cast into bars but could not be moved from the premises until it was passed and stamped by the Excise Officer. So the gold had to be stored in a strong room awaiting government inspection. The Gwynfynydd strong room door was formerly that of the old county goal at Dolgellau. After the gold had been assayed and marked by Excise it was taken to Dolgellau and despatched in bullion boxes to London.

Local folklore has it that the old copper mine dumps were seen to glow. The miners called it 'funny metal'. Actually the effect was produced by tiny specks of gold gleaming in the sunlight.

It is interesting to note that the wedding rings of Queen Mary, and Mary, Princess Royal, were made of gold mined at Gwynfynydd and that of Princess Marina of Kent came from Beddycoedwr. Appropriately the Royal St David's Golf Club's gold cup was made from gold from St David's mine.[23]

After many years, there is a renewal of interest in these deserted gold mines of North Wales. Geochemical Re-Mining has acquired a gold prospecting licence from the Crown Mineral Agents and has taken a thirty-year lease over 700 acres at the Gwynfynydd mine, and is to undertake extensive exploration and examination to mine gold, silver and lead. This mine between 1860 and 1916 produced 60,000 oz of gold worth about £500,000 sterling at the present price of gold. The company is said to have paid a hefty sum to the mineral rights owner, a former sheep farmer. The famous old Clogau (St David's) Mine is also to be reopened.

Within three miles north and north-east of Moel Fammau, the highest point of the Clwydian Range in North Wales, shafts and levels were driven in search of gold-bearing quartz, but the information is scanty.[24]

The first record we have of gold in Scotland is the grant of a title to all the gold that should accrue to David I, to the Abbey of Dunfermline in 1153. Gilbert de Moravia discovered gold at Durness in north-west Sutherland in 1245. By an Act of 26 May 1424 James I it was proclaimed that 'any mine of gold or silver found in any lord's lands of the realm and it may be proved that three ha'pence of silver may be found in a pound of lead the Lords of Parliament consent that such a mine may be the King's as is usual in their realms', indicating that previously the output of gold had been negligible.

But with the discovery of gold mines at Crawford Moor in James IV's reign (1488-1513) we find in the Treasurer's accounts for 1511 to 1513 many payments to Sir James Pettigrew for working the gold mines in that district. As early as 1513 an expert report was made to the king by John Damiane, Abbot of Tungland, whose expenses were paid by the royal exchequer. After James IV was killed at Flodden the Queen Regent controlled the mines.

In July 1526 Joachim Hoegstre, Gerard Sterk and Antony de Nikets with other Germans and Dutchmen were granted a lease of all gold, silver and other metal mines. In the following year they were granted a licence to coin. The results must have been disappointing because in 1531 there is recorded a payment to the Dutchmen, on their departure homewards, of a repatriation grant. Then in 1535 a commission appointed to look into the working of the gold mines led to the importation of miners from Lorraine in 1539.

It is fully established that James V opened up the alluvial gold deposits of Crawford Moor, five miles north-west of Leadhills, between 1538 and 1543. He won a lot of gold from the burns and rivers of these moors with the help of these French metal workers specially brought to Scotland for the purpose.

A good story is told of James V, who organized a hunt on the barren Crawford Moor in 1537. The newly arrived French courtiers from sunny France were most outspoken on the inhospitable and bleak surroundings. Whereupon the king boasted that he would provide a local product at the forthcoming banquet that would outshine the fairest fruits of France. At the banquet a covered dish was brought in, and the queen removed the cover to disclose the basin filled with newly minted gold 'bonnet pieces'.* The king won his wager. In the following four years 41¼ ounces of Scottish gold made a crown for the king, and 35 ounces one for the queen, while more gold went into numerous items of royal jewellery.

In 1567 Cornelius de Vos painter to Queen Elizabeth I, was sent by her to the Scottish Court. He went to the moorland hills of Clydesdale and 'Nydesdale' to become interested in the gold there, showing Edinburgh friends samples of the gold. He persuaded them to raise £416 and soon had six score men and women beggars working there. On one occasion he sent £450 worth of gold won in thirty days. Soon after, a Dutchman, Abraham Greybeard, got sufficient gold to make a basin big enough to hold an English gallon of liquor. King James, according to Sir Walter Scott, presented the vessel filled with gold 'bonnet pieces' to the French and Spanish ambassadors.

Bevis Bulmer can be justifiably called the first British mining engineer. Nothing is known of his antecedents but about 1578 he was engaged by an Edinburgh goldsmith, Thomas Foullis, to operate his lead mines in Lanarkshire, instead of which he devoted all his attention to gold mining.[25] He worked principally on Mannock Moor and Wanlock Water in 'Nithsdale' and on Friar's and Crawford Moors near Leadhills.

* A Scottish gold coin issued by James V in 1530, 40 Scots shillings in value, so-called from the large flat bonnet on the king's head.

He worked systematically, constructing head and tail races, diverting streams to make the beds accessible, made dams for a water supply to the buddles, and generally organized the working of these alluvial deposits on a sound engineering basis. He was fairly successful, built a large mansion at Glengonnar, kept open house, and over the lintel had carved: 'In Wanlock, Elwand and Glengonnar I won my riches and my honour.' He mentions nuggets of 5 and 6 ounces found within two feet of the moss at Langcleuch Head, where he erected a stamp mill from which he got 'much small mealy gold'. His greatest success was on Henderland Moor in Ettrick forest. Returning to England in or before 1600, he is said to have presented to Queen Elizabeth I a 'porringer of clene Scots gold', but this cannot now be traced. He was appointed in 1605 to be Chief Governor of the King's Mines, a post he held till his death in 1613.

In 1592 James VI gave the right to mine gold in Glengonnar to the same Thomas Foullis who had employed Bulmer and this right is still the property of his descendants, as Sir Robert Hope married Foullis's daughter.

In 1603 Bulmer was granted a sum of £200 to search for gold on Crawford Moor. He wanted the Scottish gold mines reopened, proposing that twenty-four gentlemen of substance subscribe £300 each to the venture and be rewarded by knighthood and called 'Knights of the Golden Mynes.'[26] This fantastic scheme was thwarted by the Secretary of State, the Earl of Salisbury much to Bulmer's disappointment. However, he himself was knighted in 1604 by James I.

In 1616 a grant was made to an Englishman, Stephen Atkinson, a refiner of the mint in the Tower of London, who tried hard to persuade James I to venture again in Scottish gold mines, but the king had lost about £3,000 in such undertakings, and was chary of another venture.

British Museum manuscripts throw light on the activities of another worthy, George Bowes, who in 1603 had been granted the right to search for gold in Crawford Moor. He gave up a year later but asserts that in the time of James V some 300 people in several summers washed gold to the value of over £100,000. This was during his own and part of the previous generation. Pennant claims that during the reigns of James IV and James V gold washed from the mountains was worth not less than £300,000. Dr Lauder Lindsay later gave the yield as £500,000. But the authority of both these claims is obscure; even George Bowes's aforesaid estimate of £100,000 is probably an overstatement.

The bleak barren moorland around Leadhills, the highest village in

Scotland, 1,300 feet above sea level, lies on the borders of Lanarkshire and Dumfriesshire. The gold-bearing area is almost entirely in rocks of lower Silurian age. The precious metal is found in streams and in gravelly clay, or 'till', disposed on the slopes of the hills. The gold is generally in an extremely fine state but with occasional nuggets; the largest, of 27 ounces, is or was in the Marquis of Linlithgow's collection.

The Leadhills district has been the centre of alluvial gold production from many named burns feeding the headwaters of the rivers Clyde and Nith since at least 1350 and gold washing had always been a significant industry in the area (see Plates 5 and 6). Bevis Bulmer was probably the most successful of the many adventurers there. Stephen Atkinson who had other links with him wrote about him in a book written in 1619, *The Discoverie and Historie of the Gold Mynes in Scotland*, but this was only published by the Bannatyne Club in 1825.[27]

Dr Lindsay in 1862 washed 975 grams for the Countess of Hopetoun, and three miners collected 33 grams from 'till' 120 feet above the bed of a stream. The late Professor J.A.S. Ritson, while Inspector of Mines for Scotland, panned enough gold to make his wife's wedding ring, which takes us well into the twentieth century.[28] However, all these were not commercial activities and according to the curator of the Dumfries Museum large-scale profitable commercial gold-mining ended by the second half of the sixteenth century, though amateurs are still panning a few grams from the burns even today.[29] A typical Leadhills gold assay was 86.60 per cent gold and 12.39 percent silver.

In Perthshire in the Breadalbane estate area above Loch Tay and in the headwaters of the river Tay a nugget weighing 2 oz was found in former times. Gold has been reported by various observers from other parts of Perthshire. There is in the Natural History Museum, South Kensington, a nugget weighing 1,010 grams from Turrerich, on the River Quaich near Loch Freuchie.

Small quantities of alluvial gold have been recorded from tributaries of the River Dee and in sea sands near Aberdeen. In Sutherland a 1½ ounce nugget is recorded as having been picked up in the Kildonan stream in 1840. Gold in any quantity was not known there before November 1868 when R. N. Gilchrist, a native, who had been a successful gold-digger for seventeen years in Australia, tried various streams for gold. His search soon discovered gold in Kildonan burn, a small tributary of the Ullie. Gold was also found in neighbouring burns. In

1869 there was a gold rush and four hundred prospectors descended on Strath Ullie. They diverted the channels of the burns to get access to the gravel in the stream bottoms. For the time being the work was rewarding, as is evidenced by the payment during the year of a licence fee for each digger of £1 a month plus a royalty of 10 per cent. Royalty was paid on £3,000 worth of gold but concealment of most of the gold was irresistible and probably the total recovery was not less than £12,000. Much of this gold was purchased by P.G. Wilson of Inverness, over £60 worth in February 1869, in April £431 worth, and by 24 August between £5,000 and £10,000 worth. When the Kildonan and Suisgill burns were exhausted the diggers spread to adjoining ground, but only got fine dust which defied their washing process. They had set up a 'gold town and traces of their work can still be seen.

Eventually the trouble suffered by the sheep farmers caused the Duke of Sutherland to take out an interdict to close the diggings,[30] which had also affected the Lower Helmsdale salmon fishing. A small team tested the ground in 1894 and in 456 days from 6 May to 6 August the following year produced only 12 oz 16 dwt. In 1911 a further survey found that it would not be worth while to rework the ground.[31]

In much smaller quantity gold was found in two streams at the head of Loch Brora. It was worked for only a short time because the licence fees did not compensate for the damage caused in driving the sheep to the bleak moorland. Digging was prohibited from 1 January 1870 and has never been resumed. A typical assay of Sutherland gold was 81.27 per cent gold and 18.47 per cent silver.

An amusing case is the excitement caused in Fife in May 1852 when an Australian convict wrote to friends in Kinnesstoun, his former home, saying rocks near his native village were like those worked for gold in Australia. Three hundred men for the better part of a month dug in a limestone quarry overlooking Loch Leven. A bed of clay contained large globular masses of iron pyrites which the diggers thought were nuggets of gold.

Assays of the gold mined in Britain show that overall about one-tenth of the bullion is silver, a little more than the world average of 90 per cent gold and 8 to 9 per cent silver. Gold mining in Britain is as ancient as the Romans but did it contribute anything to the Industrial Revolution? The considered answer must be No: NOTHING!

Cornwall and Devon's combined output can be measured only in pounds of the precious metal and was thus negligible. The Romans gutted South Wales. South Scotland ceased to be a commercial contributor centuries ago. Only North Wales remains. The Industrial Revolution created a large class of middle-class wealthy industrialists and merchants, who provided and lost most of the risk capital for the mines of Merionethshire.

Little evidence remains of gold-mining in Britain today. Naturally streams in spate have obliterated alluvial workings, except perhaps in Sutherland. Where lode gold was mined in Merioneth there is evidence of former activity in mine dumps and foundations of surface plant.

British gold needs to be refined to separate out the silver. The old wet process of parting by nitric or sulphuric acid has now fallen into disuse. Refining by chlorine gas was first developed by Miller in Sydney Mint in 1869. Gold melts at 1063 degrees centigrade, and 7000 to 8000 oz are melted in a tilting furnace and covered with borax. A pipeclay tube reaching to the bottom of the crucible introduces chlorine gas, which forms volatile chorides drawn off into the fume chamber. The last traces of copper resist chlorination until the bulk of the silver is transformed. The end point is evidenced by the reddish brown fumes of gold chloride. Chlorination is then stopped and chlorides are skimmed off and the gold is poured into bullion moulds at 996 fine. The operation takes from two to three hours.

The scum is treated for silver by agitation in a salt solution. The base metal chlorides dissolve in the brine leaving the insoluble silver chloride, which is collected and reduced with metallic iron in dilute sulphuric acid. The sponge silver is washed clean from iron salts, then melted and cast into bars.

The Wohlwill system of gold deposition developed in 1874 in Germany is still used for purifying gold in many refineries, but is restricted to bullion with limited amounts of silver because, during electrolysis, silver chloride formed in excess adheres to the anode and interferes with the dissolution of the gold. Cast bullion anodes 3 to 4 inches wide, 4 to 8 inches long and $\frac{1}{4}$ inch thick are suspended from a carrier bar in porcelain cells. These anodes are often enclosed in cloth bags to stop the insoluble silver chloride fouling the electrolyte as well as assisting its collection. The electrolyte is a solution of 4 per cent gold as chloride and 5 per cent hydrochloric acid. The high current density of 80 to 100 A/sq ft speeds up deposition and reduces the tie-up of gold. The operation takes from twenty-four to thirty-six hours and the pure sheet gold cathodes are then removed and cast into bars. The silver chloride slime and anode scrap are melted, cast into anodes and, being mostly silver, are electrolized in Moebius silver-parting cells. The process yields gold of 999 fine. The disadvantage is the amount of gold held up in the electrolyte. Because of this and the detrimental effect of a trace of lead, the process is not used as much as formerly by bullion refiners,[32] although in the United States since 1902 the electrolytic method of refinery is still favoured.

Passing chlorine gas through the molten metal was first performed by Thompson in 1838 and, as above, applied by Miller at the Sydney Mint in 1869. The refining of bullion by blowing oxygen or air through it was described by Rose of the Royal Mint.[33]

R. R. Kahan of the Perth mint in 1916 recommended passing both air and chlorine gas through the molten metal to reduce the loss of gold caused by the prolonged passing of chlorine to remove all the base metals.[34] In 1931 H. R. Hilman investigated at Perth the stage at which silver, copper and lead were removed in the chlorine-air process and found the removal of lead erratic. In 1933 Reynolds of the Melbourne mint refined an alloy consisting of 57 per cent gold and 42.5 per cent silver by chlorination.

In Britain the electrolytic refinery of Messrs Johnson Matthey was formerly the world's largest, dealing with the entire output of the Rand. The bullion was 800 to 900 fine and was refined to 996. In 1932 the electrolytic plant was replaced by a chlorination plant with a capacity of 400,000 ounces a week.

When the Royal Mint moved from the Tower of London in 1810 provision for a refinery was made. This, known as the Royal Mint Refinery, was leased to Rothschilds in the mid nineteenth century and sold to them in 1924. Before World War I the sulphuric acid and electrolytic processes were used, but at the end of the war a chlorination plant was installed to bring rough gold from 800 to 992 fineness; only the last stage to 996 fineness was electrolytic. Continuously working, it has a capacity of refining 400,000 ounces a week.[35]

CHAPTER THREE

Silver

Professor John Percy, 'the Father of English Metallurgy', wrote: 'It is a remarkable fact that silver should invariably be present in galena (sulphide of lead) sometimes in very minute proportions usually in the form of sulphide.' This interesting personality studied in Paris under the great chemist Gay-Lussac. At his father's behest, he studied medicine. On graduating at Edinburgh University he was appointed physician to Queen's Hospital, Birmingham, where he found himself at the very centre of metallurgical interests. In 1839 he married his cousin, an heiress, and could then follow his own inclinations without financial anxiety. He became lecturer in metallurgy at the recently established Royal School of Mines, later becoming Professor. In 1849 he was made a Fellow of the Royal Society.[1]

Silver was the other precious or noble metal produced in Britain, so logically this chapter follows that on gold. The mining of silver also comes under the Mines Royal. Almost all the silver from Britain has come from silver-lead mines, thus some readers may prefer to read the chapter on lead first.

Nevertheless, a little silver was won from the Cornish copper mines. At Wheal Cock near St Just in 1753 a piece of native silver as big as a walnut was found. Fibrous native silver associated with horn silver (chloride) was met with in and above the adit on 1788 to the value of £2,000 at Wheal Mexico near Perranzabuloe. At Herland in Gwinear argentite (sulphide of silver) was found in a crosscourse near the intersection of a copper lode, as well as pyrargyrite (sulphide of silver and antimony) and black oxide, more than half silver, called by the miners 'gooze dung' ore. £8,000 worth was won, some of it smelted at the mine and the rest sold at Bristol.[2] Dolcoath, near Camborne, produced native silver[3] and silver ores from the 160 fathom level. £3,000 worth

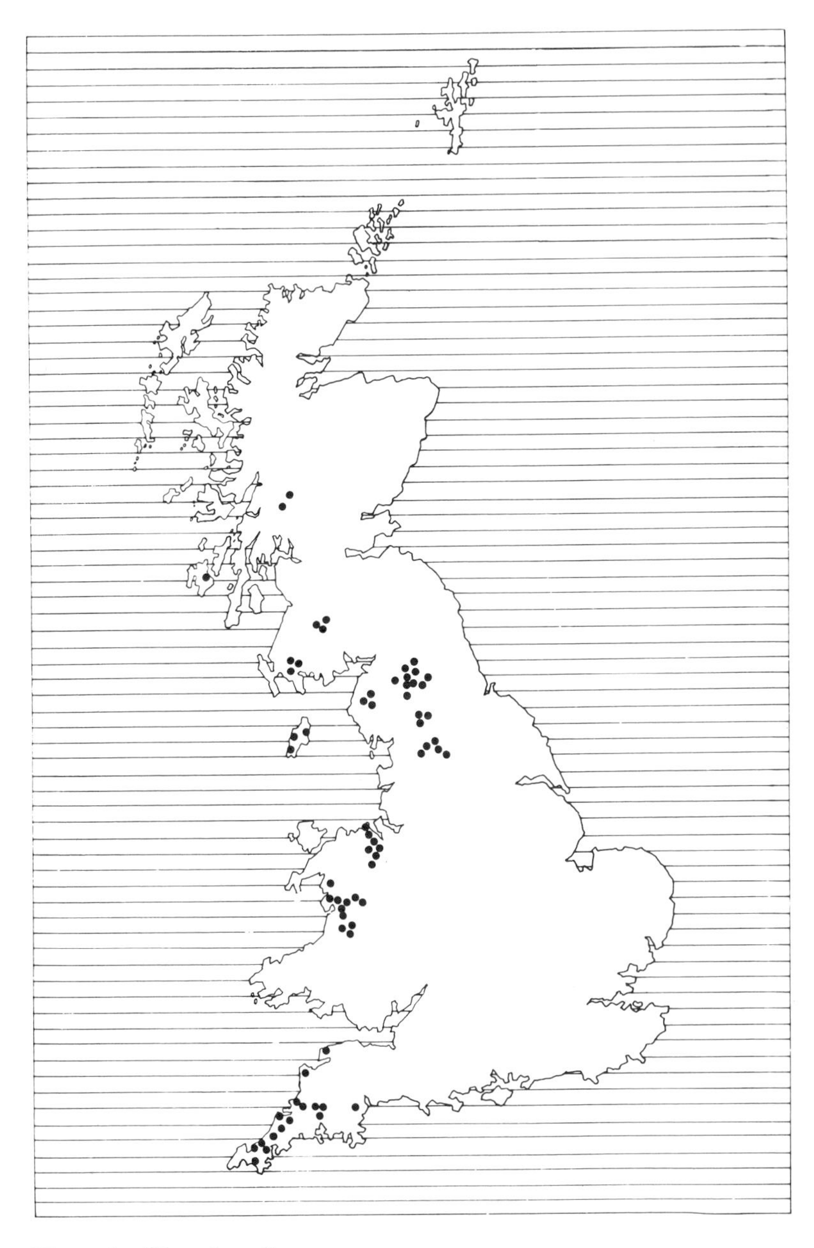

Figure 3 Silver deposits

was quickly won in 1833, and, because he generously relinquished large amounts of dues in troublesome times, Lord de Dunstanville was presented with a service of plate made from Dolcoath silver. At Wheal Duchy near Calstock in east Cornwall native silver, ruby silver, a sulphide of silver and antimony, and grey and black sulphides were found in a cross-course trending north-east to south-west in 1812, and from these ores a silver cup was made and presented to the Duke of Cornwall. In 1816, £6,000 worth of silver was taken from a leader only four inches wide in a lode a foot or less in width. Some lumps in it were 70 per cent silver.[4] In 1833 the mine was reopened as Wheal Brothers. From the silver-bearing gozzan, ore was sold varying in price from £2 to £500 a ton. From the Cornish copper mines in all perhaps over 10,000 ounces troy were produced, a tiny proportion, however, of the many millions of ounces won from lead mines.

The silver-bearing lead mines occur in sedimentary rocks in the form of veins, as replacements and disseminations in limestones and dolomites (magnesium limestones) of Carboniferous age and in older rocks of Silurian and Ordovician age often as steep-dipping orebodies. The nearer the surface, the higher the silver content. Weathering, the breakdown of rocks by temperature changes, the freezing of water, root growth, abrasion and the solvent action of water containing oxygen, carbon dioxide and organic acids from decomposed vegetation causes carbonates, sulphides and chlorides of the metals present to form. Beneath this, but still in the oxidization zone, the deposition of these metallic oxides and carbonates occurs. Then there is the upper sulphide zone with the special feature of enrichment due to the chemical interaction and deposition of sulphides and lastly the sulphide deposit in its primary state. Silver does, however, migrate but is more readily precipitated than lead so there is the tendency to silver enrichment in the upper sulphide zone. This is sometimes a most prominent feature and is the explanation of the origin of many bonanzas.[5]

The Romans were lucky in working a number of lead deposits that had not been developed before in many parts of the province and thus mined the silver-rich upper portions of the orebodies. We know from the inscription on Roman pigs of lead marked 'Ex: Arg:' that the lead metal had been desilvered. Unfortunately we have no records of the quantity of silver produced by the Romans in Britain, but historians think it must have been considerable and that much of it was sent to Rome to make coins for use throughout the Empire.

We know little about the production of silver after the Romans departed, except that the practice of desilvering lead bullion declined. In England silver plates and other vessels were first used by Wilfred, a Northumbrian bishop, 'a lofty and ambitious man', in A.D. 709.[6] It might have been British silver because that corner of England produced a great deal in later centuries.

Desilvering was in practice again before the Norman Conquest. Little is learnt about silver from Doomsday Book but towards the end of the thirteenth century two groups of rich silver mines were discovered and both were in Devon forty miles apart: Combe Martin a few miles east of Ilfracombe on the north Devon coast and the Bere mines on the sheltered peninsula formed by the confluence of the rivers Tamar and Tavy. No one knows when and by whom they were discovered, but Combe Martin is thought to have been found by 'tinners' searching for alluvial tin and the Bere mines by labourers digging up stones with a metallic sheen. In both cases the value of the find was quickly realized and the king, Edward I (1272-1307), claimed ownership of all metalliferrous mines by medieval custom rather than by any well-defined prerogative and seized the newly won silver for his treasury. Thus these two groups of mines helped to finance the warlike enterprises of the Plantagenet kings.

The earliest record dates from 1293 when William de Wymyneham, King Edward's factotum, accounted to the Treasury for 270 lb of refined silver, 521½ lb in 1294, and 704 lb in 1295. Between 12 August and the end of October 1294, 370 lb of silver ore were shipped to the king from Maristow on the river Tavy and by 1297 silver worth £4,046, and £360 worth of lead, were sent to the Royal Treasury. The ore can only have come then from shallow pits, trenches or opencast workings.

As early as 1295 the king under royal warrant impressed 340 miners from the Derbyshire Peak and 25 from Wales. In 1360 a writ was issued authorizing certain persons 'to arrest and imprison such as should resist till they should give security to serve the king in the said mines'.[7] The system of impressment of miners for the Devonshire mines continued throughout the fourteenth century. Little is known of these mines for over two centuries after that date, but in 1485 over a thousand men were employed at Combe Martin and the Bere mines together. Some of the mines on the two parallel lodes in the Bere peninsula yielded between 80 and 100 ounces of silver to the ton of ore, and even up to 140 ounces to the ton was recorded.

After it had been lying unworked for at least half a century Adrian Gilbert and John Poppler, a Londoner, in 1587 discovered a rich new vein at Combe Martin. They sold a half share to the same Bevis Bulmer who was active in the Leadhills alluvial gold workings. He was to pay the cost of mining and smelting and to have half the ore. The venture, though successful, lasted only four years and made profits of over £6,000 for each partner. With the last of the silver metal extracted, Bulmer had two fine silver cups made by a London goldsmith called Medley. One of the cups weighing 137 ounces was presented to Sir Richard Martyn, Lord Mayor of London. This great silver bowl was melted down and made into three tankards which still remain in the Mansion House collection of plate and are inscribed 'The Gift of Bevis Bulmer'.

Thomas Bushell, the protégé of Sir Francis Bacon, came into the picture during the Civil War. He had secured the post of Warden and Master of the Mint at His Majesty's Mines Royal in the Principality of Wales. He shifted to Shrewsbury, on to Bristol and then to north Devon where at Combe Martin he minted many silver coins. After a few years' activity at the end of the eighteenth century, in 1835 the Combe Martin and North Devon Mining Company was formed with a capital of £30,000. Half the capital was spent on shaft sinking and machinery and good ore was reached and some dividends paid. Between 1845 and 1847 silver-lead ore valued at £65,000 was won.[8] The claim in 1864 that employment was assured for centuries was not fulfilled.

In the small peninsula around Bere Alston, only about ten square miles in area, there are two strong parallel lead lodes running north to south and about three-quarters of a mile apart. On the westerly lode the mines from the north were Ward, North Hooe and South Hooe. On the longer easterly lode they were, from the north, Buttspill, Lockridge, Furzhill, East Tamar Consols and South Tamar Consols.

In Edward I's reign, in 1299, ore from these mines was pledged to Florentine financiers, the Frescobaldi. It turned out to be a bad bargain for them and after six years it reverted to the king and then gave a record amount of £1,775 worth of silver and over £810 of lead.

Because of the king's demands and the high wages paid, experienced European miners were attracted and this highly skilled labour resulted in greatly improved techniques. Most important was the introduction of 'avidots', or adits, self-draining levels which doubled the output as the work could be carried on in winter as well as in summer. For a few years after 1485 the king is said to have got £44,000 a year from these mines, but by 1500 increasing difficulty in dewatering had led to their abandonment.

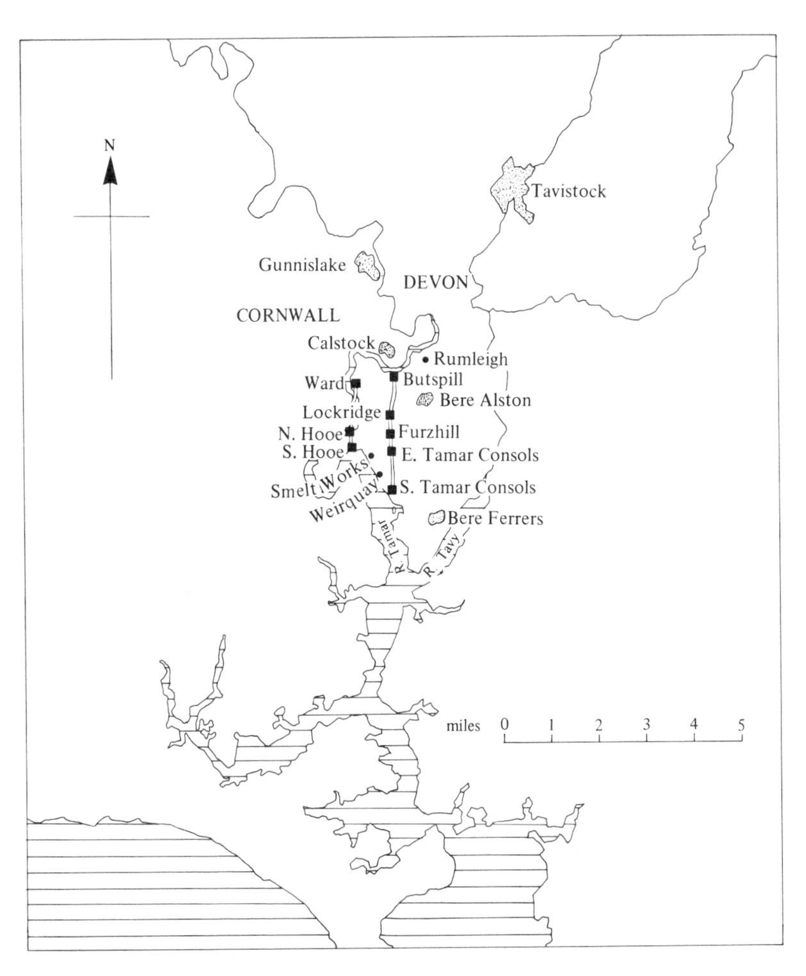

Figure 4 Bere Alston silver-lead mines

Timber came down the river Tamar in barges and also refuse from tanneries to form the absorbent bed of the furnaces for extracting the silver from the bullion by cupellation. As smelting was done on the spot, the forced draught was by bellows worked by a waterwheel, and the furnaces for reducing the ore resembled blacksmiths' forges.

For a couple of centuries until the beginning of the nineteenth century these Bere mines sink into obscurity. However, between 1784 and 1785 South Hooe mine produced 6,500 ounces of silver from ore as rich as

140 ounces of silver to the ton. By 1809 the Bere mines again became leading producers of silver and lead in the kingdom. Steam power helped in the recovery and for the next forty years a thousand people were employed. The rich pockets of silver-lead ore under the Tamar were attacked. The two principal mines were South Hooe on the western lode and South Tamar Consols on the eastern lode.

South Hooe was moribund in 1835 when it was acquired by the Tamar Silver Lead Company and Percival Norton Johnson was put in charge. He introduced a number of important improvements, fan ventilation, a 25-degree inclined shaft for both easy haulage and a walkway for the miners in lieu of the toil of climbing ladders. The ore in the deepest levels still yielded between 50 and 60 ounces of silver to the ton.

In the concentration plant Johnson introduced shaking tables of German design. By 1852 nearly £44,000 in dividends was paid out on an original capital of £12,000. South Hooe and North Hooe were worked together as Tamar Consols. The ore was smelted and the silver extracted at Weir Quay and these works were bought and re-equipped by Johnson. Its eighteen furnaces could treat over 300 tons of ore a month. The quay was deepened to take ships of 400 tons burthen and silver lead ores from Spain, France, Newfoundland and Wales were sent there for treatment.

In 1850 Johnson introduced the Pattinson process detailed later in this chapter allowing ore containing as little as 3 ounces to the ton of ore to be successfully treated whereas previously not less than 8 ounces silver to a ton of metal was considered profitable. Johnson's humane enlightened attitude to his workpeople was for that time unusually good and kind. He retired early from the Tamar mines and died at Stoke Fleming in 1866, a lonely, disillusioned and enigmatic person. The mines and smelters that he managed were never so prosperous after his departure. However, from 1845 to 1876 they produced 326,300 ounces of silver and more than 9,000 tons of lead. With the output of previous centuries this property produced well over 600,000 ounces of silver and was the richest of the Bere group of mines.

The South Tamar Consols was the richest of the eastern group, and its flooding in 1856 at 8 p.m. on 30 August, luckily on a Sunday when no workmen were underground, ended its life and that of the adjoining East Tamar Consols whose workings were connected with it.[9] The only visual evidence today are forgotten quays, old chimney stacks and grass-grown dumps of a group of mines that produced over a million ounces of silver in the last hundred years of their lives.

In South Devon in the Upper Teign valley a group of three mines yielded 10 to 20 ounces of silver to the ton but ceased production in 1880. The Frank Mills mine was the deepest of these and the most important.

In East Cornwall three miles west of the river Tamar between Calstock and Callington there were seventeen mines working in 1850. Between 1852 and 1909 the Hingston Down mines produced over 33,000 ounces of silver and nearly 9,000 tons of lead. All these mines had shut down by the end of the century. The Mary Ann Trelawny lode four miles to the west was very rich in silver and was exploited to over a mile in length. The ore produced was worth nearly £1 million sterling. Wheal Trelawny was worked from 1844 to 1871 and Wheal Mary Ann from 1846 to 1876. Their dressed ore carried 30 to 90 ounces of silver to the ton. Together they once employed a thousand workpeople and on a capital of only a few thousand pounds made profits of £130,000. Parallel and 360 feet to the east, a lode on which Ludcott mine was situated was remarkable in producing native silver, sulphide of silver and the combined sulphide of silver, arsenic and antimony in considerable quantities. Herodsfoot mine 3½ miles south-west of Liskeard was in parts rich in silver and made a profit of £5,000 a year.

In North Cornwall there are numerous veins of lead ore at small intervals all the way from Tintagel to St Agnes with some very ancient workings. The old Treburgett mine had remarkable occasional shoots of ore rich in antimonial silver ore. At Garras mine three miles north of Truro the lode was from 2 to 6 feet in width and heaved twice by slides. The richest silver ore was between these slides, some of the lead ore yielding 100 ounces of silver to the ton. The mine reopened in 1814, the ore being smelted at the mine and the silver extracted on the spot. The metal contained 70 ounces of silver to the ton. The veins were found by Sir Christopher Hawkins when draining a marsh.

Wheal Pool near the head of the Looe Pool was working at the end of the sixteenth century and was working again profitably in the mid-eighteenth century. In 1790, after being idle, it was reopened when the ore was found to contain from 30 to 60 ounces of silver a ton.

According to the mineral statistics from 1845 to 1886, when lead mining practically ceased in the West of England, Cornish lead ore had yielded 5,721,153 ounces of silver and Devon gave 1,341,597 ounces, a total of 7,062,750 ounces. Although in early years the figures are avowedly defective, it is probably safe to give a total of 7½ million ounces of silver.

In Cornwall and Devon, however, silver-lead mining had been pursued for 600 years previously, perhaps longer, and some of the mines had been worked extensively and, near the outcrops, were rich in silver. Although there were long periods of mining inactivity, it is considered safe to assume the six centuries' output to be as great as the forty-one years mentioned above, bringing the total of silver to 15 million ounces from workings totalling not more than 20 miles long with an average width of 4 feet and a depth of about 600 feet.[10]

The rich silver-lead mines of Cardiganshire were rediscovered, drained and operated by Hugh Myddleton with great success for fourteen years from 1617. His fame however rests on the New River scheme by which he brought water to London from springs in Hertfordshire; for this work and the reclamation of land in the Isle of Wight he was created a baronet in 1622. The same Thomas Bushell mentioned above, came to terms with Myddleton's widow and obtained a grant of the Mines Royal in Cardiganshire. He dewatered these mines by a long crossout adit which he ventilated by pipes and bellows, a method long used in Germany but applied here successfully for the first time in Britain. He was helped by Joseph Houghstetter, the son of Daniel. Against much opposition he also established a mint at Aberystwyth. Bushell wrote several books, the most important being *A Just and True Remonstrance of the Mines Royal in Wales* in 1641. His extraordinary career is told in a biography by J. W. Gough, entitled *The Superlative Prodigal* and published by Bristol University.

In mid-Wales according to Professor O. T. Jones the hundred or more mines clustered around Plynlimon together produced 3¼ million ounces of silver. In addition there was the silver produced there that made the fortunes of Hugh Myddleton, Thomas Bushell and others. In less than forty years, 1827 to 1866, Goginan mine yielded 497,000 ounces of silver and Van mine contributed over 750,000 ounces. So these two mines alone produced 1¼ million ounces of silver metal.

From 112 tons of metal smelted from ore from the Bwlcy-y-Plwn mine in Merioneth 2,150 ounces of silver were extracted. Seven small mines in the St Tudwal's peninsula yielded 13,520 ounces of silver; in Anglesey the complex lead zinc ore called 'bluestone' between 1882 and 1911 yielded 84,426 ounces. In the Llanrwst area lying between Conway and Bettws-y-Coed in Denbighshire the Parc mine's galena contained silver, and Gorlan mine produced from 621 tons of dressed ore 2,233 ounces; the richest mine in this district was the Trecastell mine three miles south-west of Conway which from 1892 to 1913 yielded 74,875 ounces.

During the reign of Charles II the London Lead or 'Quaker' Company was formed and its life extended from 1692 to 1905. Its history is extraordinary and worth studying. From 1692 to 1704 it was concerned with the Royal Mines Copper and the Ryton Company which produced silver. From 1704 to 1710 was a period of consolidation into the London Lead Company. From 1710 to 1790 was the period of its widest expansion in all parts of the British Isles searching for silver-rich lead ores. From 1790 to 1882 the Company confined its activity to the counties of Durham, Cumberland, Westmorland, Northumberland and North Yorkshire, forming one big group of silver-lead mines. From 1882 to 1905 came its decline and reduced activities.

The Ryton Company had been formed before 1696 and in 1697 was buying lead ore from Alston Moor for smelting and extracting silver at its works on the river Tyne. There was a close connection between the Ryton group and the Welsh company, for in 1703 a site was secured at Gadlis near Bagillt on the river Dee where a smelting house was established with four furnaces for smelting lead ore, a refining furnace for silver and a slag hearth. This plant was supplied by ore from mines on Halkyn Mountain, leases of which had been secured for forty years.

On their tours of lead markets, members of the London Lead Company's Court heard of good silver-lead ore for sale and arranged visits to the mines. They examined mines in Blanchland in the Derwent valley, Teesdale, Wanlockhead, Stirling, the Isle of Man, the Mendips, Derbyshire, South Ireland and as far as the Orkneys, leasing some of the mines. Remembering the remoteness of mining districts in those days, with all land journeys made on horseback, one must have great respect for those pioneers. After the successes at Ryton and Gadlis the Company preferred to lease and work mines rather than buy ore.

Before the London Lead Company confined its activities to the Northern counties, their smelter at Gadlis near Bagillt using Welsh ores produced 430,604 ounces of pure silver between 1704 to 1790, according to Pennant's history of Holywell dated 1796, and their Ryton smelter near Newcastle in the first forty years of its history produced over a million ounces. The richest ore smelted by them was from the Clargill vein near Tyne Head with over 36 ounces of silver to the ton of lead metal. Other veins gave 12 to 16 ounces but the general average was only 8 ounces to the ton.

In company minutes concerning shipments of silver from Wales they reported shipwrecks, attempts at salvage on the wild coasts of Pembroke and Gower, piracy and chases by French privateers.

Gadlis closed down about 1790 after ninety-eight years. Mines scattered in Denbighshire, Merioneth and Cardigan were gradually given up and in the case of leases centred at Halkyn, the sources of ore were confined to within thirty miles of the Gadlis mill. The principal leases were from Mr Mostyn, and a small colliery near Gadlis leased from Mr Pennant supplied all the fuel for the smelting plant and the domestic needs of the workpeople. More and more water in the Halkyn mines increased pumping costs and a loss of £9,099 was made between 1787 and 1789. So the mill and stock at Gadlis were sold for £700 and the Halkyn mine leases surrendered.[11]

Before that, in 1721, the Alva silver mine up the Silver Glen half a mile east of Alva, near Stirling, came within the company's reach. In 1711 Sir John Erskine discovered the vein and brought miners from Leadhills to work it. On opening up the mine a large mass of ore containing native silver was discovered: 12 ounces of pure metal to 14 ounces of ore. The weekly output averaged £400 and Sir John gained up to £50,000 but this rich ore was soon exhausted and the silver ore gave place to copper ore. This mine was forfeited by Sir John Erskine on his joining the Pretender. He was pardoned through the influence of the Tsar of Russia, a relative of his, and the mine was restored to him. The company surrendered the lease in 1831. Sir Isaac Newton had commissioned Dr Justus Brandshagen to inspect the Alva mine. He reported that the silver ore was the best in any part of Europe.[12]

In 1782 the London Lead Company leased and developed the Great Laxey mine on the Isle of Man which it later sold to the Duke of Atholl for £1,250. At Foxdale, also in the Isle of Man, 2,882,440 ounces of silver were produced from 1845 to 1910. In addition silver was won from the galena mined at Great Laxey, elsewhere on the island and in earlier times. Great Laxey mine was operated more or less continuously from 1700 to 1919, the deepest shaft being over 1,800 feet deep.[13] In 1876, 2,500 tons of lead ore from this mine yielded 103,332 ounces of silver or over 40 ounces to the ton. The total silver yield for over two centuries is not recorded, but must have been enormous. In the south-west of the island the outstanding outcrop at Bradda Head was rich in silver.

In 1790 the London Lead Company sold all its Derbyshire leases. From its mines on Alston Moor, and in Teesdale, Weardale and Westmorland, where it concentrated its energies, in 1737 it extracted from 630 tons of lead metal 7,706 ounces of silver. In 1760 from 2,040 tons of lead metal 10,992 ounces of silver were extracted. During the nineteenth century the following yields were achieved:[14]

	Ounces of silver
1825	39,862
1828	35,000
1851	58,493
1867	55,155
1871	47,874

The silver content was higher in the shallow depths which were worked in the early period of the major leases and at Hudgill Burn and Greengill the oxidized ores were much richer in silver than the mines working in sulphide ores.

In 1709 the Company showed an interest in Scottish mines. An agent was sent from London and miners from Newcastle to examine mines on the mainland and Hoy in the Orkneys and to the Isle of Stronsa, where Earl Morton had a silver mine. In 1710 the Company was involved in the Wanlockhead silver-lead mines and began to operate smelt mills and a refinery, as well as building houses and making roads to Leith for the carriage of lead and silver for sale and shipment. Long and deep drainage adits were made. The miners there formed a Friendly Mining Society and in 1721 proposed amalgamation with the Company. For several years they cooperated and in 1727 ended their association in a friendly manner. All mine leases were surrendered in 1731 and the mill and equipment handed over to the Friendly Mining Society which prospered for several more years.

Contemporaneously a silver-lead company was founded by Sir William Blackett in 1699, a family concern handed on through his successors to the Beaumont Company, which however did not become a limited liability company until the last four years of its existence. It began to operate in the Allendale valley. Early in the eighteenth century it acquired mines in Weardale from the Bishop of Durham's estates. Fifty mines were operated by it in south Northumberland and Durham. The Weardale leases were surrendered in 1883 after nearly two centuries of productive mining. Starting at much the same time as the London Lead Company they came to an end together, though the London Lead Company continued on a small scale at Teesdale and Bollihope until winding up in 1905; whereas the Beaumont Company, as W. B. Lead Mines Ltd, continued until after World War I.

The combined output of silver of the two companies was:[15]

	Years	*Lead concentrates*	*Ounces of silver*	*Ounces of silver per ton lead concentrate*
Beaumont Co.	1729-1870	1,072,697		
Beaumont Co.	1725-1870		3,045,000	2,8
London Lead Co.	1815-1870	469,487	2,405,000	5,1
Totals		**1,542,184**	**5,450,000**	**3,5**

The London Lead Co. additionally produced one million ounces of silver from 1704 to 1745. Dr Raistrick adds to the London Lead Company's 1815-70 figure a run of twenty-six years from 1740 to 1765 during which 270,821 ounces of silver were produced. So that the two companies together were responsible for 6¾ million ounces of silver. At Cargill Head Mine in March 1748 1½ cwt of lead yielded 1,600 ounces of silver. Maximum yields were over 50 ounces and 30 ounces of silver per ton of metal were common.

The Commissioners of Greenwich Hospital in 1735 were granted the estates of the Earl of Derwentwater which had been sequestrated because of his participation in the Jacobite Revolt of 1715. They included the Manor of Alston Moor with its lead mines. Most of the latter were leased to Colonel George Liddell who failed to exploit them profitably. The 'Quaker' Company gradually took over his leases as they fell vacant. The Commissioners had their own smelt mill at Langley, Northumberland, from 1768. Attracted by the increasing prosperity of the mines, they erected their new mill near Langley Castle in the Tyne Valley, convenient for receiving the ore from Alston Moor and also for sending the metal to Newcastle for sale and shipment. This mill was built with three hearths, a slag hearth, two refining furnaces and a reducing furnace. In addition several peat houses were built. The whole cost was £1,400. Two waterwheels were made for the mill and refinery and a dam to hold three days' supply of water constructed. It was found necessary to increase the water supply however and a new cut was made in 1774 to store more water in the dam. The problem of fuel was a vital one: additional peat houses were built and coal from local collieries bought. Experiments were made with charcoal but these were abandoned and purchases of peat and coal increased. The project was successful from the start. From 1769 to 1774, from 4,204½ tons of lead 33,507 ounces of silver were produced. The sale of silver was recorded in great detail. The Commissioners were excellent employers and housed and paid their workers well. They also paid compensation for losses to local farmers from the death of cattle caused by the sulphurous fumes emitted from the smelt-house chimneys.[16]

At Leadhills and Wanlockhead lead mines in 1873 produced 10,720 ounces of silver. This group of mines had produced silver during many centuries and according to official records from 1842 to 1920 Wanlockhead produced 520,000 ounces of silver and Leadhills 314,380 ounces. During the two previous centuries with no continuous records available much silver-rich lead ore was won. In those earlier days much of the metal was exported to be desilvered. So it seems safe to assume a total of 1¾ million ounces of silver from these two rich Scottish mines.

Other lead mines in Scotland were silver-bearing. The Bellsgrove mine in Argyllshire yielded over 4,000 ounces of silver between 1847 and 1871 and at Tyndrum in Perthshire the steel ore assayed 40 ounces of silver to the ton of ore but the ordinary galena only yielded 10 ounces.

At Hilderstone, two and a half miles NNE from Bathgate Station in Linlithgowshire in 1606 a collier named Sandy Maund found a heavy red ore full of white threads of metal which Bevis Bulmer proved to be silver. He worked the vein in collaboration with the owner, Sir Thomas Hamilton of Binney. The ore was native silver associated with niccolite. It is said to have yielded 480 ounces of silver to the ton worth £120, the native silver fetching 4*s* 6*d* an ounce. In 1870 a shaft was sunk to 225 feet, from the bottom of which a borehole was sunk a further 360 feet, but only volcanic ash was found. In 1896 the mine was reopened and worked for galena but was soon abandoned.

Robert Hunt, the keeper of Mining Records at the Museum of Practical Geology recorded that from 1857 to 1859 the lead ore at Grassfield mine in Cumberland contained 30 ounces of silver to the ton. The Frank Mills mine in Devon from 1859 to 1866 averaged 26 ounces to the ton. Among Cornish mines, Wheal Mary from 1857 to 1864 averaged 65 ounces of silver to the ton of ore; Treveathen mine 70 ounces to the ton from 1854 to 1859 and Trelawney mine 57 ounces to the ton of ore. At Foxdale in the Isle of Man, two lots gave 12 to 20 ounces of silver to the ton of lead and 59 to 80 ounces to the ton.

When there was a revival of lead mining in the 1930s, from primary sulphide ore the silver content at Mill Close Mine in Derbyshire was as low as 1 to 1½ ounces—too little to merit extraction. During the same period at Halkyn it was 2 to 4 ounces, but as all the lead concentrates were exported to the Continent it is not known whether any silver was extracted from them.

As the highest output of lead in Britain was recorded in 1856, the table below may be taken as representative of the industry at its most prosperous. It is interesting to note that the lead mining fields that had been worked intensively by the Romans, the Mendips, Shropshire and Derbyshire, yielded little silver.

Silver produced in Great Britain in 1856

	Mines operating	Dressed ore *Tons*	Silver *Ounces*
Cornwall	37	9,974	248,436
Devonshire	13	3,138	77,456
Somerset	1	750	—
Shropshire	9	4,408	—
Derbyshire*	39	9,534	—
Yorkshire	14	12,174	302
Westmorland	11	2,924	23,860
Cumberland	73	7,311	51,931
Durham and Northumberland	34	24,125	79,924
Carmarthen	2	1,280	—
Cardigan	38	8,560	38,751
Radnor	2	13	—
Montgomery	13	1,724	2,660
Merioneth	4	349	572
Denbigh	2	3,104	1,034
Flint	39	4,607	19,340
Caernarvon	5	237	—
Isle of Man	3	3,218	60,382
Scotland	8	1,931	5,289
Sundries (production of under 10 tons / year)		161	550
Total	347	99,514	611,488

* Records do not give the names or number of operating mines in Derbyshire in 1856. In a later year, there were 194 mines, only 39 producing over 10 tons a year (Phillips Report, 1918, p. 157).

During a somewhat extended period of the Industrial Revolution well over 20 million ounces of silver are recorded as being produced. How much more was unrecorded is not easy to estimate, but it is clear that silver was then, as it had been for many centuries previously, an important product and a great asset to the economy of the country. The probable total output of silver in this island, including the Isle of Man, from Roman times cannot have been far short of 60 million ounces.

1 Roman adit at Ogofau Gold mine, Pumpsaint showing tapered entrance to allow miners to carry out broken ore on their shoulders in baskets or skin sacks.

2 The stone at Pumpsaint, Carmarthenshire used in Roman times for breaking down the ore.

3 The entrance to St David's Clogau Gold Mine, Merioneth, *c*. 1891.

4 Panning for alluvial gold near Leadhills, Dumfries.

5 Buttspill Silvermine engine house and ivy-covered chimney stack, Bere Alston Peninsula, Devon.

6 Morwellham Quay, Bere Alston Peninsula. The reed grown patch in the middle distance is the site of the dock dug out for the shipment of silver, copper and arsenic from 1859 onwards. Note the granite bollards.

7 Ancient circular settling, or buddle-pit, near Geevor Tin Mine, Cornwall.

8 A convex buddle at Dolcoath Mine, Cornwall.

9 The engine house and shaft headgear at South Crofty Mine, Cornwall.

10 Wheal Jane Tin Mine, near Truro. The first tin mine in Britain to come into production for over half a century. This shows surface installations including two headframes and the concentrator in the background.

11 Nineteenth-century engine houses of tin mines which ran below the sea, Botallack, Cornwall.

12 LydfordCastle, near Okehampton, Devon. This was used as a stannary prison in the early days of Dartmoor tin mining.

13 A tin streamer's works in the Portreath Valley, Cornwall.

14 Hemispherical liquation pot-furnace for refining impure tin *c.* 1925.

15 An entrance to Britain's oldest bronze age copper mine, Stormy Point Mine, Alderley Edge, Cheshire.

16 Parys Mountain, one of the largest opencasts in Europe. On the summit of this west pit are the engine house, chimney and great spoil banks.

The extraction of silver from lead bullion for thousands of years was accomplished by cupellation. Not only was lead bullion so treated but lead metal was added to all silver ores to extract the pure silver metal.

In a reverberatory furnace a dished hearth of absorbent material was made of bone ash, finely ground calcined bones, for choice; though in medieval times charred tan turves were in use in Devon and Derbyshire and marl elsewhere.

The bullion and fuel were laid on the hearth and with a blast of free air the metal was heated to a temperature of 900 to 1000° Centigrade. The bullion melted and the lead was oxidized to yellow litharge, protoxide of lead. Litharge, from the Greek, literally means stone of silver because of its use in the refining of that metal.

The molten litharge was absorbed by the porous bed, any excess flowing over the rim of the 'test', or bed, to be collected and solidified in iron receptacles. The litharge melted well below the temperature at which silver is oxidized. So the molten silver metal remained at the bottom of the furnace bed and on cooling solidified into a solid cake of silver.

The furnace bed of bone ash, marl or tannery turves was removed and, together with solid litharge overflow, was recovered as lead metal in a reducing furnace.

The 'test' or hearth bottom of absorbent material does not become corroded as would brickwork or stone, and with good design and skilled operation the loss in lead was small. The treatment of lead bullion on a large scale is in a large reverberatory furnace with a shallow concave bed of suitable absorbent material for the retention of the silver capsule. Continuous blast is achieved by the blowpipe being machine operated. When heated beyond its melting point, silver is not oxidized by atmospheric air either when alone or in contact with such a highly oxidizing agent as litharge, but other metals present are speedily oxidized and quickly dissolved in the molten lead oxide leaving the silver free from impurities.[17]

Cupellation must have been successfully developed at least a century before 2,500 B.C. since pure silver artefacts appeared at that date as far apart as Ur, Troy and Egypt. The people of Pontos, a part of Cappadocia on the south-eastern corner of the Black Sea, a country backed by the Pontine mountains and adjoining mineral-rich Transcaucasia, are credited with this discovery which greatly enhanced the popularity of silver and inevitably increased the production of lead metal especially in Asia Minor about the middle of the third millennium B.C.[18]

Cupellation is referred to in the Old Testament in the Book of Jeremiah, chapter VI, verses 29 and 30: 'The bellows are burned, the lead is consumed of the fire, the founder melteth in vain for the wicked are not plucked away. Reprobate (or refuse) silver shall men call them because the Lord has rejected them.' In this passage all the essential points are mentioned, the artificial blast, the oxidation of the lead and the silver dross, reprobate silver or litharge.

The Romans in Britain preferred bone ash if they could get enough of it. Their normal silver metal in the first century A.D. was 99 per cent silver, the remainder being mostly lead. Ancient Britons apparently did not cupel as their lead contained much more silver than Roman lead.

The pundits tell us that 'Ex Arg' on Roman lead pigs must be expanded to read 'Ex Argentariis' that is 'from the silver works' and not 'without silver'. But does this matter, as obviously the lead sent to the silver works was to be desilvered? A list of eleven Roman pigs assayed gives an average silver content of 2 ounces 10 dwt of silver. Good practice indeed for cupellation as until the Pattinson process was invented it was not thought worth while to desilver lead containing less than 8 ounces of silver. Admittedly labour and material costs were very different in Roman times. After the Romans left, desilvering of lead bullion declined and Saxon lead contained ten times the amount of silver as that from the Romano-British refineries at Silchester.

To come to much more recent times, the London Lead or 'Quaker' Company excelled over its rivals in good design and the high skill of its furnacemen for the extraction and refining of silver. It was the best in this country, in fact in Europe, as a silver refiner. The Company found that the bone ash they were purchasing in London was impure, so they bought bones and made their own. After a time they made the curious discovery that bone ash when mixed with stale beer was greatly improved, so they added small brew houses to their smelting and refining establishments.

In the early years of the London Lead Company, from 1705 to1737, all its silver was sent to the Royal Mint where it was used in the redemption of the coinage. In consideration of this the Company's Court petitioned the Crown for a device to be stamped on all coins made from its silver. Sir Isaac Newton, then Master of the Mint, took up the plea with the Treasury and after some delay in 1706 Queen Anne gave warrant for such a device. This was renewed in 1709 by correspondence between Lord Godolphin and Newton, and the device was used on most issues of five shilling, two shilling, one shilling and sixpenny coins of Queen Anne and Kings George I and II.

The silver delivered to the Mint was cast into ingots and after assay the value of such ingots was credited to the Company's Treasurer. Without exception the silver gained a large extra payment as it proved on assay to be considerably finer than the Mint's standard silver. Throughout its life the London Lead Company maintained its tradition of superiority in the extraction and refining of silver. A recent chief Assayer of the Royal Mint from 1926 to 1938, Dr S.W. Smith, remarked that the Company's silver was in those earlier days sought after for its high quality and suitability for fabrication. It was never a matter of sentiment but was because of the silver's superior ductility resulting from such successful refining.[19] The Company was selling its silver to the Mint at less than marked price and wished to be entirely free to sell on the open market. But the solicitors of the Mint in 1766 claimed that the 'Quaker' Company was bound to sell to the Mint. The Company stuck to its decision that nothing in its Charter obliged it to sell at 53/2d an ounce troy. After much deliberation the Mint's solicitors agreed. Nevertheless for long after 1766 Quaker silver was still bought by the Mint for coining Maundy Money. The recorded output of silver for the whole of Britain was:

in 1855 — 561,906 ounces of silver
in 1856 — 610,488 ounces of silver
in 1857 — 532,866 ounces of silver
in 1865 — 724,856 ounces of silver[20]

most of this from the northern smelters and refineries of the Quaker and the Beaumont Companies.

Hugh Lee Pattinson's new process revolutionized the recovery and extraction of silver from silver-lead bullion. He was born at Alston in Cumberland in 1796 of humble parents and when he grew up was employed as a lead assayer by the Commissioners of Greenwich Hospital at Langley in 1825, and he also worked for both the London Lead and Beaumont Companies. He was self-made, self-educated and ingenious. On 28th October 1833 he patented 'An improved method of separating silver from lead', and read a paper on the subject before the

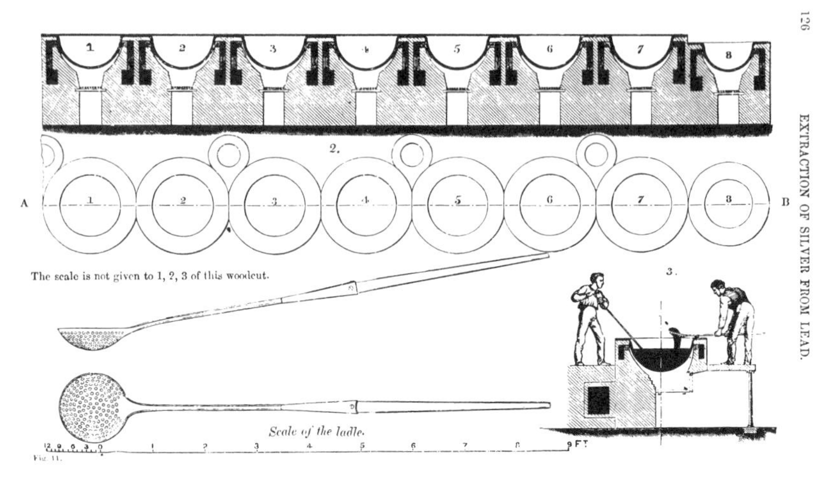

Figure 5 Pattinson process for the extraction of silver

British Association.[21] By his method the molten lead is slowly cooled to permit lead crystals lean in silver to form, leaving below a silver-rich liquid. The lead crystals are freed from metallic impurities just as water in freezing is freed from saline matter dissolved therein. As the first crystals contain less silver than the remaining fluid, by repeated strainings of the crystals the silver is concentrated in the remaining fluid up to 200 or even 300 ounces a ton of metal; beyond that concentration the process ceases to be payable.

The work was carried out in a series of large hemispherical iron pots with a capacity of 6 to 8 tons of bullion, and the crystals separated and collected on large perforated ladles. The process of ladling, melting and crystallizing continued until the lead ladled out contained under 2 ounces of silver to the ton of metal.

John Henry, manager of the lead smelting works at Gadlis, used a line of eight pots, with perforated ladles, the pots holding up to 8 tons of molten metal. Each pot had its own fire-box for the coal fire to melt and remelt the metal.

John Taylor in 1850 confirmed Pattinson's claim of 15*s* a ton of bullion, to cover the cost of desilvering, loss of lead and the expenses of desilvering and refining the silver.[22] The silver enriched bullion was then cupelled in the way detailed above.

Pattinson gained a reward of £16,000 for his invention, small enough compared with the wealth it conferred on others. One North Country smelter whose annual income was increased by £5,000 begrudged the inventor his small share of the profits. Pattinson died at Newcastle upon Tyne on 11 November 1858. The London Lead or Quaker Company was always in the forefront of any technical advance in the production of silver and purchased the right to use the Pattinson process for a thousand guineas. It was tried out first at their Nenthead smelter. In 1837 the Company reported on the great advantages of the process, noting that only 1/373rd part of the lead used was lost in the process. The Company's agent, Robert Stagg, was responsible for highly satisfactory improvements involving some changes in the Pattinson process. In 1839 it was reported that Stagg's inventions saved a great deal of the labour hitherto employed in it. This improved method of desilvering was adopted at the Company's refineries at Stanhope, Bollihope, Eggleston and, as aforesaid, Nenthead, and remained in use there until the end of the Company's activities. Pattinson and Stagg remained good friends throughout their association.

The Ryton smelter was sold to the Beaumont Company by the London Lead Company, where with the Pattinson process bullion containing 6 ounces of silver was economically refined whereas formerly it would never have been considered for extraction. Johnson, down on the Tamar in south Devon as aforesaid, also quietly adopted the Pattinson process and was able profitably to smelt lower grade ore.[23]

Alexander Parkes of Birmingham invented a process patented in 1850 for the recovery of gold and silver from lead bullion. Usually zinc metal from 1 to 2 per cent of the weight of the lead was added to molten bullion and heated to the melting point of zinc, which combined with the precious metals and separated out as a scum consisting of zinc, all the silver, and some lead. This was strained off and separated. The scum was distilled in retorts, leaving the silver with a little lead. It was then cupelled.

By this process the proportion of silver in the remaining lead may be reduced to half an ounce per ton of metal or even less. Silver from 400 to 500 ounces was contained in one hundredweight of zinc, separated by dissolving in hydrochloric acid, dilute sulphuric acid or volatilizing. In 1851 a second improved process was patented by Parkes. His process by 1870, superseded Pattinson's to a great extent and it is now almost universal with considerable advantages over Pattinson's, as it produces a product for cupellation far richer in silver and at a lower cost. Pattinson's process with its repeated meltings and crystallizations is conducive to higher losses in lead and silver. In Parkes's process lead with 50 ounces of silver requires 1.3 per cent zinc to the ton of bullion, and lead with 200 ounces of silver requires 2.0 per cent zinc to the ton of bullion. The zinc is introduced in two operations to attract all the silver. The molten mass is stirred well with a perforated rabble.[24]

If 1750 is taken as the approximate date of the commencement of the Industrial Revolution, then much silver was produced in this country before that date but the records show that millions of ounces were put on the market during its duration and the two major inventions for improving the separation of the metal occurred towards the end of it.

The accumulation of wealth created by the revolution greatly increased the demand for jewellery and for many kinds of domestic vessels, utensils and tableware in solid silver or silver plate. A Sheffield cutler, Thomas Boulsover, in 1742 discovered the process of coating thick copper with thin silver by fusion. Joseph Hancock developed it from 1750 to 1765 and thereafter there was rapid growth in the production of Sheffield plated articles. Robert Adam and John Flaxman later inspired the makers to the creation of much more intricate designs. The excellent appearance and cheapness caused a widespread demand, but by 1865, at the end of the revolution, the industry was dead. It had been killed by electrolytically deposited silver.[25]

The so-called nickel, or German silver, contains no silver at all but as a base when silver plating was successfully launched was an alloy of 50 per cent copper, 25 per cent zinc and 25 per cent nickel, hence E.P.N.S. (electro-plated nickel silver) on tableware.

William Lewis and T. Wedgewood advanced the art of photography which used silver nitrate and answered the demand for pictures by the new wealthy middle class. Silver was consumed by advances in dentistry, by the electrical industry, in solders for brazing and for conductors, thus increasing the demand and consumption of silver and the alloys and salts derived from it. Silver in an annealed condition is the most perfect known conductor of both heat and electricity.

The archaeological remains of smelters and refineries are intimately bound up with the lead industry described in a later chapter.

CHAPTER FOUR

Tin

In 926 Athelstan banished the West Welsh from Exeter and made the river Tamar their eastern boundary. In 937 he conquered Cornwall and some years later, in 951, Edmund Duke of Cornwall granted some sort of charter to the tin miners of Devon and Cornwall. No one can tell how far back the stannary* customs originated nor whether, as tradition has it, the 'tinners' of both Devon and Cornwall held a joint parliament on Hingston Downs just over the Tamar. It is not until 1156 that the documentary history of the tin mines began when the small production was mostly confined to west Devon and even then the written evidence was little more than a list of figures from the pipe rolls.

Until 1198 the sole connection between the tin industry and the Crown was the collection of an annual tax with the occasional use of the right of purchase. In 1194 Richard the Lionheart sailed away to the French wars from Portsmouth leaving Hubert Walter, Archbishop of Canterbury, in charge of the country's administration. To make money for his master, and appreciating the value of the tin works of Devon and Cornwall, he placed them under the supervision of William de Wrotham as Warden.

The output of tin improved after the Norman Conquest but fell to 400 thousand weight† in 1200 so perhaps King John, in granting the first charter to the stannaries in 1201, had his personal interest at heart. His charter confirmed the old privileges of

> digging tin and turfs for smelting it at all times, freely and peaceably and without hindrance from any man, everywhere in moors and in the fees of bishops, abbots and counts...and of buying faggots to

* The word stannary is derived from *stannum*, the Latin word for tin settled as such in the fourth century A.D. So stannaries simply means tin mines.

† A thousand weight was a technical term meaning 1200 lbs.

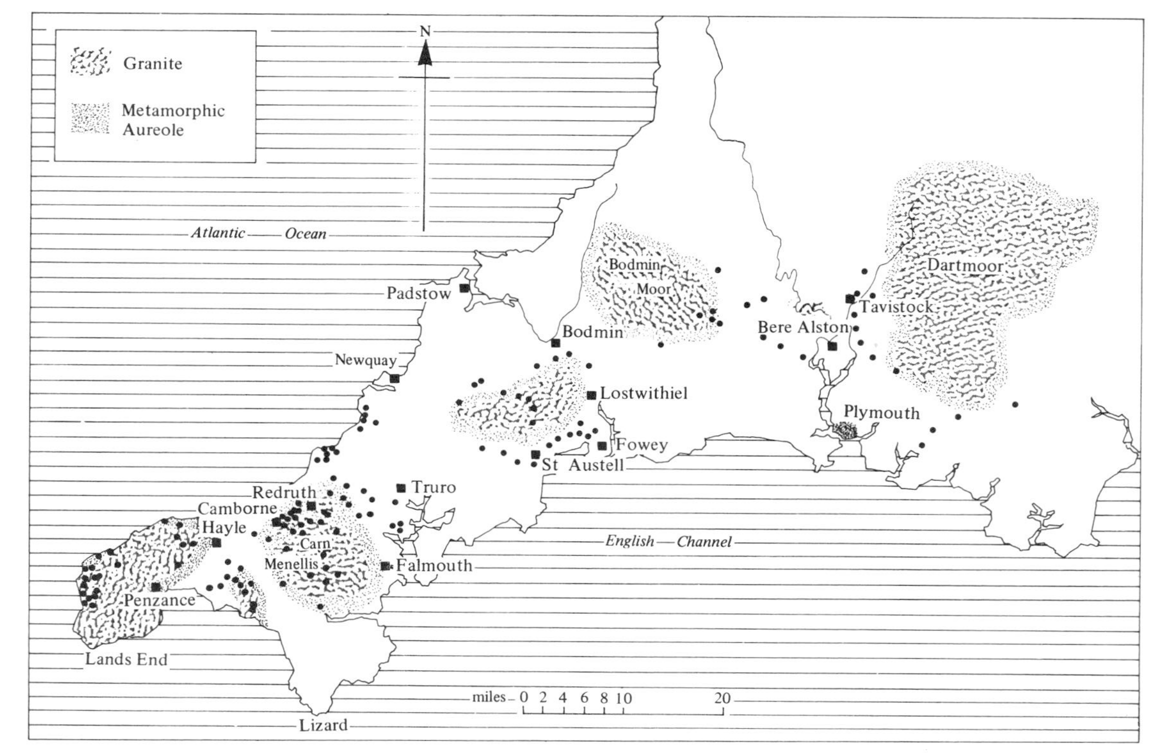

Figure 6 Tin mines of west Devon and Cornwall, showing the five major granite domes

smelt the tin without waste of forest and of diverting streams for their works and in the stannaries just as by ancient usage they have been wont to do.[1]

At the end of the thirteenth century the Cornish tinners petitioned the king to restore their charter of liberties. So Edward I issued and granted a new charter in 1305 with features which remained the real constitution of the stannaries, freedom on the king's estates for the working miners with the privilege of being solely liable to the warden and his officers for any actions arising in the stannaries except for those involving land, life or limb. The existing stream works were grouped into stannaries with their separate courts. In Devon, Chagford, Ashburton, Tavistock and, later, Plympton encircled the Dartmoor stream works. The moorland between Launceston and Bodmin where the river Fowey rises became the stannary of Foweymore. The area around Hensborough Beacon, Roche, Luxullian and St Austell became Blackmore, divided into eight tithings. According to old chronicles the tinners' charter was kept in the tower of Luxullian with eight locks and eight keys. A small area from the north coast to Truro became the stannary of Tywarnhayle. The two stannaries of Penwith and Kerrier north of Helston and between Lelant and Land's End were the last. The boundaries were never accurately defined.[2] King Edward I's charter also confirmed the already existing tin coinage or stampage. All tin metal had to be taken to a coinage town, taxed and thus made free for sale. Coinage produced considerable revenue for the Crown but, as England in the Middle Ages had a virtual tin monopoly, in the end these taxes were paid by the consumer.

Coinage was a matter of considerable ceremony in the coinage towns. Liskeard and Lostwithiel for east Cornwall and Truro and Helston for west Cornwall. In course of time, as production steadily increased in the west, Penzance was added as a fifth coinage town in 1663. To one of these towns all tin metal had to be brought twice, and later four times a year. Coinages held only at Midsummer and Michaelmas bore hardly on the working tinner, who could thus only sell his tin twice a year and had to borrow money to live and support his family. The lender, in most cases a merchant or middleman, usually charged excessive rates of interest. So Henry VII decreed two extra coinages because 'the poor tinners have not been able to keep their tin for a good price when there are only two'.

At noon on the first day of the coinage the Royal Officers, the controller and receiver with hammer and weights, the weigher and assayer, porters, merchants, pewterers, factors, with a few Italian and Flemish traders assembled as all the tinners produced their tin. The blocks of tin were brought in by the porters to be weighed one at a time; the weight was called out by the 'peaser' when it was taken over by the assayer who rapidly assayed it to ensure that the metal was of good quality. If so it was struck with the Duchy Arms to signify that it was pure and that coinage on it, a tax of four shillings a hundredweight, had been paid. According to the quantity of metal waiting to be coined the ceremony might last up to twelve days.

The whole object of coinage was, of course, to obtain the maximum sum for the Crown as well as to prevent the illicit sale and smuggling of tin out of the country. A stranger would have been much struck by the sight of blocks of tin metal lying about the streets of a coinage town in heaps, each block worth ten guineas but weighing nearly three hundredweight and not easy to steal.[3]

Cromwell abolished the coinages altogether and during the ten years of the Commonwealth the Cornish tin industry boomed. The monarchy reintroduced coinage and the same difficulties arose. Long before coinage was finally abolished it had become a futile ceremony, delaying the sale and purchase of tin, and during the centuries the collection had become so elaborate that necessary expense over and above the 4*s* tax amounted to 1*s* 3*d* a hundredweight because there were fees to cutters, to poisers, to pilers, to scalesmen, fees for using the beam, as well as drink money and dinners to the officers of the coinage.[4] So in 1838 the whole system was quietly swept away and instead a small excise duty levied at the smelting house. This, let it be noted, only towards the end of the Industrial Revolution.

The king's taxation of the tinners was not capricious but usage hardened into law, in general it was lighter than for other classes, thus causing all sorts of people often only remotely connected with the industry to claim its privileges. This led to frequent commissions of *oyer* and *terminer* (to hear and determine) set to ferret out wealthy men becoming 'tinners' to escape other taxes. Up to 1335 they were directly answerable to the king, but after that they came under the Prince of Wales, as Duke of Cornwall.

The wardens and vice-wardens appointed stewards who administered the Stannary Courts which were held every three weeks, when they dealt with every action of debt or trespass for both black and white tin and seized smuggled tin, imprisoning malefactors in the special stannary prisons at Lidford in Devon or Lostwithiel in Cornwall. Lidford prison got the unenviable reputation of hanging malefactors first and trying them afterwards—if found innocent a priest prayed for their souls.[5] The stewards acted as judges in their respective districts. In cases where one of the persons involved was not a 'tinner' it was still tried in the Stannary Courts.

The Cornish tinners' parliament met behind closed doors, either at Truro or Lostwithiel, to elect a speaker who opened the session, and members then proceeded to debate means of redressing or amending any abuses within the stannaries. Until 1636 the enactions were signed by the warden, vice-warden, the duke and all twenty-four stannators, but after that year only sixteen needed to sign. The session might last several weeks. The Lord Warden suggested to the king that the members should be sober and loyal persons and, as the nomination of representatives was the privilege of the mayors and councils of the stannary towns, they were usually owners of large mines or tin dealers and were largely chosen from the chief families of Cornwall. Perusal of lists showed few who were not baronets, knights, esquires or gentlemen. So the mayors became the ordinary tinners' bitterest enemies.

The Devon parliament on the other hand was chosen quite differently as twenty-four 'jurates' from the four stannary districts of Chagford, Tavistock, Ashburton and Plympton included all classes of tinners and the ninety-six chosen representatives met at Crockerntor, which was once an assembly site of the Ancient British. It was a much more democratic gathering. At the two leet courts held each year the stannary officials in open court in the presence of a jury took solemn oath to fulfil their offices justly and well.

The unlawful practice of suing for debts direct to the vice-wardens undermined the authority of the lower courts in such cases as debts for goods or by tinners for wages. For this reason the lord warden, vice-wardens and stewards declined to hold any courts until the passing of the Stannary Act of 1837. Subsequent statutes on the stannaries took the bold step of abolishing stewards' courts entirely, restoring common law. The vice-wardens had to be barristers holding court in Truro at least once in three months. The Stannaries Courts Abolition Act of 1896 transferred all jurisdiction to the County Courts. So after 700 years the Stannaries Courts and customs came to an end.

The supply of tin was a Cornish monopoly throughout the Middle Ages. It was exported by several foreign routes, French, Dutch and Venetian to the Eastern Mediterranean Basin until towards the end of Queen Elizabeth I's reign when the Levant and East India Companies carried Cornish tin in British ships.

The Crown benefited greatly financially from the tin industry and Richard Lionheart even stooped to gamble in the tin trade. Smuggling, to avoid the onerous coinage tax, was rife from de Wrotham's time and at its worst in the reign of James I when most Dutch tin merchants possessed a stamp with the Duchy arms to make the smuggled tin pass for the best English brand. Government inspectors were powerless against a whole population of born smugglers.

From the earliest records of tin output in Britain in the twelfth century the rich deposits of West Devon around Dartmoor produced nearly all European tin up to the end of the thirteenth century, an average of 70 tons from 1156 to 1160 rising to almost 350 tons by 1189. Not until the beginning of the fourteenth century did Cornwall overtake Devon's output. Ever after, Cornwall's production was the greater, averaging some 550 tons annually up to the mid-seventeenth century. At the beginning of Cornwall's ascendancy East Cornwall was the main producer, but the trend in output was always westward.

There was an outburst of mining activity up to the Black Death which struck England in 1348. In 1337 there was a record output of 715 tons, but then the plague almost ruined the stannaries. So great was the demoralization that the Black Prince, as Duke of Cornwall, in 1351 gave orders that on pain of forfeiture the tinners must spend the same labour and money as before the Black Death because in that year only 265 tons was produced—not much more than a third of 1337's output.

By the middle of the sixteenth century output rose to over 800 tons but fell to little over 500 tons by the end of it. In the first half of the seventeenth century it grew to 700 tons a year and to between 1,000 and 1,500 tons in the second half, influenced by the growing importance of lode mining. In the first half of the eighteenth century it advanced to 2,000 tons per annum. In 1719 Cornwall produced 2,150 tons and Devon only 30. In the second half of that century the output grew to 3,500 tons. By the first half of the nineteenth century production increased to about 5,000 tons, and the second half included the tin boom. The ten yearly averages during the second half of the nineteenth century were:[6]

	Tons metal	
1851 - 1860	6,400	
1861 - 1870	9,403	
1871 - 1880	9,654	
1881 - 1890	9,237	
1891 - 1900	6,405	the beginning of the decline
1901 - 1910	4,583	

The decline was rapid. In 1893, towards the end of the boom, 14,000 tons of black tin were produced but in 1896 only 7,663 tons.

There were only 75 tin mines in 1801 but 200 by 1838, a figure which had fallen to 180 by 1859. Williams's *Mining Directory* gave 340 mines in 1862, but probably this included stream works and perhaps stamps, because a government report gives the maximum number as 230 in 1874.[7] This fell to 100 in 1900, to 30 in 1918 mostly in the Redruth to Camborne area, to four mines in 1938 and only two in 1949. Today there are the same two mines, Geevor and South Crofty but there is now a third new one, Wheal Jane near Truro. During the boom the major contributors were the mines of the Redruth-Camborne district. Dolcoath, once the largest and deepest tin mine in the world, produced from 1896 to 1903 an average of over two thousand tons of concentrates annually.

In 1865 Cornwall produced 40 per cent of the world's tin[8] but in ten years it had dropped to less than a quarter and by 1939 to one per cent while today it is negligible.

The price of tin metal was nearly £6 a ton in 1199, rose in the middle of the seventeenth century to £65 and remained, with some ups and downs, the same a century later. Yet by the middle of the nineteenth century it was over £100 a ton and by 1913, £198. At the end of 1972 it was well over £1,500 a ton, and in February 1974 reached the unprecedented price of over £3,525 a ton.

Tinstone, cassiterite, binoxide of tin, SnO_2, was the first mineral to be mined and reduced to metal in Britain thousands of years ago. Recent excavations at Treviskey revealed a hoard of tinstone dated by associated pottery as late Bronze Age.[9] The British deposits of tin, moreover, were and are strictly confined to Cornwall and west Devon, no tin is found elsewhere in Britain.

It happens that tinstone, or black tin, possesses physical and chemical

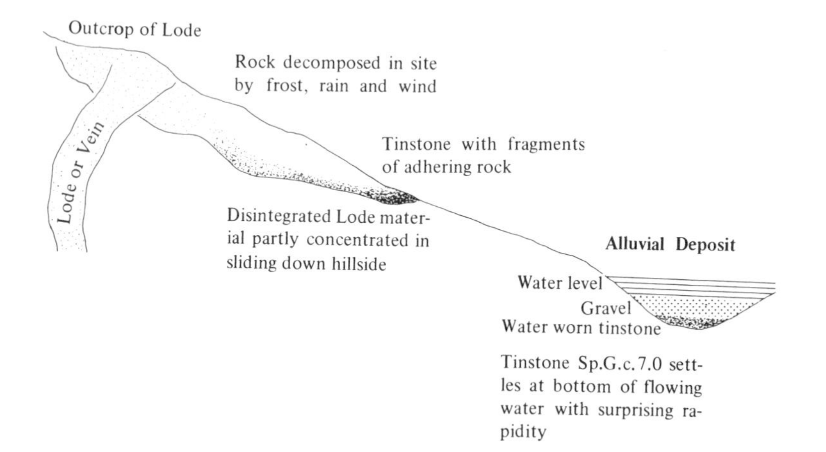

Figure 7 Eluvial deposit

properties that enable it to withstand weathering to such a degree that it remains unaltered through geological ages. Even in the oldest tin-bearing gravels the mineral is in the same state as originally formed. In the ancient stream beds of Cornwall and Devon it has been concentrated by natural agencies because its specific gravity is from 6.4 to 7.1.[10]

Where did the detrital, eluvial* or stream tin come from? According to geologists the tinstone of Devon and Cornwall occurred in lodes, that is the infillings of fissures in pegmatite dykes known locally as elvans in the granite and the adjacent metamorphosed† sedimentary rocks surrounding the granite, known locally as killas.

The five large and several smaller granite masses were part of the great hercynian mountain system thrown up many millions of years ago, and were once high mountains worn down by perpetual weathering to the domes they are today. The fissures in them and the auriole of surrounding killas were similarly eroded, the tinstone contents freed by disintegration of the associated rock *in situ*, partly concentrated by the removal of the lighter material by frost, rain and wind, settled in a wide range of sizes, but to much of it fragments of rock adhered. Some of

* Ore-bearing gravels resulting from the disintegration *in situ* of the rocks which contribute to their formation.
† Altered by heat and pressure.

these beds were deposited many million years ago, when the land surface was many feet below the present one, and were covered by the further weathering of the mountains. Nevertheless, occasionally in recent times tinstone has been found near or at the surface. At Bunny, near Botallack, a deposit at the surface was discovered, according to tradition, by a horse kicking up tinstone on the road, and no less than seven separate beds of tinstone were worked there.

Naturally the first workers, the Iberians, exploited the tin ore at or near the surface. The ancient tinners used a form of ground sluicing to win the tinstone from the eluvial deposits formed in past geological times. Their tools were primitive, at first wooden and much later wooden shovels tipped with iron, iron picks and wooden bowls. Small groups of men and boys formed partnerships to divert streams of essential water to their place of work.

The medieval tinners' oldest and most valuable privilege was 'bounding'. The discoverer put three turfs at each corner of the plot he wished to claim, or poles with a furze bush on top. Within that plot no other man could search for tin, so that whilst bounding prevailed the poorest villein could become his own master by laying out a claim and registering its boundaries at the nearest Stannary Court. The Devon tinners could enter any man's land, enclosed or not, and dig for tin. But the Cornish tinners were prohibited from entering enclosed land without the owner's leave. Moreover if they found mineral there the owner could work it, farm it or leave it unworked. On all unenclosed land, however, the Cornish tinner could search freely. An omission in stannary law made no provision regulating the amount of land included in a pair of bounds, so that in 1786 the whole of Dartmoor, all 50,000 acres of it, was claimed by a single prospector.

When a streamer took a lease from a landowner and agreed to pay him a part of the clean tin concentrate produced it was generally agreed to be from one sixth to one ninth of the output as settled between them. Alternatively the tinner paid the landowner 20*s* to 30*s* a year for each labourer employed by him. The bounder, not the mineral lord, had the right to grant licences to work the ground on terms settled between him and the adventurers. A 'dish' of one-sixteenth in the eighteenth century became one-fifteenth by 1830, equally divided between the bounder and the mineral lord.

As always, the bounds had to be registered in the steward's book at the nearest Stannary Court. When in the eighteenth century underground or lode mining came to the fore bounding became excessively complicated; to open a large lode mine entailed reconciliation between numerous mineral owners and many bounders each with their petty rights. This led to long delays, and many lawsuits in the Stannary Courts, a bountiful harvest for the lawyers.

Originally, under the 1305 charter, the 'tinner' was only a manual labourer at tin works and only as long as he worked there. But Henry VII's Charter of Pardon 1507 included gentlemen bounders, owners of tin works and blowing houses and buyers of both black and white tin. Even before that, and because the privileges were considerable, tinners had come to include streamers, smelters, carriers, colliers (that is makers and transporters of charcoal), carpenters, smiths, tin merchants, even owners of tin bounds; in short, anyone engaged in getting and preparing tin for the market. 'Tinners' were exempt from military service, except that in wartime the Lord Warden could levy troops and act as their commander. When Sir Walter Raleigh was Lord Warden, because of their expertise in digging and trenching, tin miners were employed on the defences of Plymouth Citadel in 1597.[11] Although exempt from tolls and dues at fairs, ports and markets and also tithes, under stannary laws they were like soldiers under military law.

As the centuries marched on, so the recovery of the tinstone became more laborious as more overburden had to be removed; sometimes as much as 40 feet or more covered the tinstone layer, when a small shaft or 'hatch' was sunk down to the tinstone. If the deposit was deemed rich enough a start was made at the lowest point of the valley to open a trench or level to carry off the water with the waste soil as the workings became progressively deeper.

Throughout the ages dewatering became more and more difficult. It was ladled out with a scoop, a windlass and bucket, a hand pump, a rag and chain pump and in more recent times a waterwheel with bobs.[12] However, apart from progress in methods of dewatering these surface workings, the winning of eluvial tinstone remained primitive and unchanged over the centuries. A development of the windlass and bucket was the horse whim, where a rope from the 'hatch' passed round the barrel of a huge upright drum or cage revolved by a team of horses.

When the level was clear of water each shovelful of tin-bearing gravel was thrown over an inclined plane of wooden boards, a primitive buddle, and washed with a cascade of water, the gravel being turned over and over with a shovel till the waste tailings were washed down-stream leaving the precious ore, about one half tinstone, still to be further concentrated. The ore was then raised to the surface of the moor where it was cleansed in a small launder called a 'gounce'. The finest sizes were treated by passing them through a sieve of wire or horsehair called a 'dilluer', an operation needing much skill in manipulation. After two washings, the richer product was ready for sending to the blowing house. The middlings had to be crushed and separated from associated rock. This was done in a 'crazing mill', which was a pair of grindstones, the upper stone perforated with four holes symmetrically arranged around a central larger hole for inserting the ore to be ground and the smaller holes for iron bars to revolve the upper stone by muscle power. It was similar to the Roman hand-mill. For even bigger pieces large stone mortars were used with a heavy round stone or perhaps an iron tool as pestle.

The most famous of surface workings was the Carnon valley north of Falmouth where the ancient river bed rich in tinstone was below sea level. The tinners worked knee deep in water and a dam had to be built to keep back the tide. A party of streamers in 1812 on Drift Moor near Penzance, in what was thought to be virgin ground, found an almost completely wooden rag and chain pump 6 feet below the surface, and in 1852 the head of a 'hatch' (exploratory shaft), sand-filled and made of square-framed mortised oak black with age was discovered 10 feet below the surface.

Century after century streamers washed their tailings down the valleys into the rivers, naturally causing damage to arable land and silting up harbours formerly accessible to shipping. In the Hayle River in North Cornwall the tide formerly flowed right up to St Erth's Bridge and in the Par Estuary, the tide used to reach St Blazey Bridge. Henry VIII brought in laws to prevent indiscriminate washing causing disastrous silting, but these statutes of 1532 and 1536 had little effect.

During the fifteenth and sixteenth centuries a considerable number of German miners and ore-dressers were introduced into England under royal patronage because of their superior knowledge of mining and metallurgy. In the latter century a number of them came to Cornwall. About 1580 Sir Francis Godolphin, an early mine owner of great importance in Cornwall, brought in Burchard Cranyce (Krunz) to improve the work at Godolphin. He probably erected the first stamp mill in Cornwall and is reputed to have re-treated successfully tin-bearing residues left by the Cornish miners.[13] In the first chapter of thisbook reference was made to Cornish names of German origin. Eldred Knapp, who died on 16 February 1956, was the last of a long line of German mineral dressers who came to Cornwall in Tudor times and continued throughout nearly four centuries to practise their art. Incidentally the title 'captain', given to all Cornish mine foremen for centuries, was almost certainly introduced by the Germans who came to work in the county in the sixteenth century. In Germany the mine foremen (captains) carried special staffs and candleholders underground which may still be seen in continental mining museums. They also had a special uniform in which they paraded to church on Sunday in the 'free' mining cities of Central Europe.[14]

Lode-mining started in a small way during the fifteenth century on lodes that outcropped and were worked from the surface down to 60 feet or more. Such workings were called 'goffens' or 'coffins'. Carne Seahole was half a mile long in 1822. Another attack was on exposed veins in the cliffs of the north coast. Near Perranporth and St Agnes such workings produced a veritable rabbit warren of openings.

As lode mines grew deeper the time spent by the miners climbing down and up ladders grew longer and the task more tiring. To overcome the exhausting and time-wasting task of climbing up a thousand feet of nearly vertical ladders at the end of the shift, in 1841 the Royal Cornwall Polytechnic Society offered a prize for the best machine to bring miners to the surface. Michael Loam was awarded the prize and his invention was installed at Tresavean mine shortly afterwards. This was originally powered by waterwheel but was converted to steam power in 1843. This 'man-engine' was double acting rotative with a 36 inch cylinder, with 5 to 1 reduction gearing connecting it to the two large rack-and-pinion gears which actuated the two vertical rods extending down the shaft. The men stood on small platforms attached to these rods. At every stroke they were raised 12 feet by stepping alternately from one rod to the other. If the 'man-engine' was a double one the men soon reached the surface. However, by 1862 only eight had been installed in the whole of Cornwall, and in 1881 H. M. Inspector of Mines described the attitude of the mining companies as 'disgraceful supineness'. But in many cases the old shafts were as crooked as a donkey's hind leg and too narrow to allow the installation of a 'man-engine.' In some mines where it was installed there was no single shaft that reached the bottom. For instance at Dolcoath the bottom of the mine was at 362 fathoms, but the 'man-engine' only went down to 240 fathoms. In 1886-87 there were still only ladders in West Wheal Seton and miners working at the bottom levels had to climb 266 fathoms to the surface, nearly 1,600 feet. At Levant, the mine beneath the sea, on Monday, 29 October 1919 the 'man-engine' collapsed killing thirty-one of the miners travelling on it.[15] By the end of the century most if not all the larger mines had installed 'gigs' or cages to bring the men up to surface and take them down to work.

Richard Trevithick, born at Illogan near Camborne on 13 May 1771, was responsible for many inventions. In 1813 at Hayle foundry, owned by the Harveys to whom he was related, he invented a hand-operated boring machine. The early boring machines were massive, rail-mounted units with stretcher bars to steady them and it took some time to set up and dismantle them. Between 1850 and 1875 some sixty patents were taken out on boring machines. The first serious Cornish trials were about 1860 when Tincroft mine took a twelve-month option on a Doering machine. It was not until 1861 that percussive rock drills operated by compressed air at 112 p.s.i. were available. There was a revival of interest in the 1870s, and in 1875 Dolcoath had a rockdrill that had to be brought to surface for repairs every day. In 1886 Cooks Kitchen, Carn Brea and Dolcoath all had rockdrills and the last in the same year ordered a new compressor to operate eight to ten machines tentatively for stoping as well as in development headings.[16] In the early days piston drills shared the field with hammer drills, but with improvements in the latter they began to supplant piston drills at the turn of the century.[17] Though the release of compressed air improved the ventilation at the faces the rock dust created by dry drilling markedly increased the incidence of miners' phthisis from about 1890. However, by 1920 hammer drills were so modified and improved, with hollow steel and water fed through the orifice to remove cuttings, that the dust hazard was eliminated.[18] After World War I there were great advances with wet drilling and improved blasting control.

Nevertheless, during the first decade of the twentieth century in many British metal mines the miners, in stoping, still drilled holes with moil and hammer rotating the chisel-ended drill at every blow to make a circular hole to fit the cartridges of explosives. In development faces, on the other hand, bar-mounted compressed air-operated drifters were in common use. Right up to World War I many British metal mines were not fully mechanized.

For dewatering the mines the famous old Cornish beam engines were still in use and candles were the only method of underground lighting.

They were stuck with clay on to the Cornish hard hats which were in general use long before such hats were employed in other British mining fields. Candles were not replaced by acetylene cap lamps until a year or two before World War I. The blanket-lined canvas coat and canvas trousers made up the uniform of the Cornish miner. In those days also 'drys' for drying the wet clothes were installed in all the larger mines.

In 1902 the Home Secretary asked the eminent scientist J. S. Haldane to investigate the serious occurrence of anaemia among underground workers in Dolcoath mine. The cases of so-called Dolcoath anaemia admitted to the miners' convalescent home grew from one in 1893 to twenty-nine in 1897. Accompanied by H. M. Inspector of Mines, Haldane found the mine air, a possible cause, to be on the whole unusually pure. After research he found the cause to be 'hook worm'—ankylostomiasis—and proved infection by the skin as well as by the mouth. This endemic disease is common among miners in hot countries. The Latin poet Ovid (48 B.C.—A.D. 18) noted the pale anaemic miners of southern Spain. It probably came to Cornwall from a 'Cousin Jack' returning from mining in the tropics. Haldane published a learned paper in June 1903 which was widely discussed.[19]

Whether this has any relation to the sensible and hygienic introduction of the crowst bag is not known, but in the middle of the first decade of this century it was in universal use. It was a small cotton bag with a string at the top to close it completely, enveloping the food eaten underground. The bag was slowly retracted whilst the food was eaten, thus preventing the fingers touching the food and inhibiting the spread of 'miners' worm'. Normally the food was a Cornish pasty.

One of the most distressing sights in the early years of this century was the miners home from abroad dying from 'miners' phthisis' caused by the inhalation of dust created by dry-drilling underground. A goverment inquiry about 1900 showed the death rate of Cornish miners from twenty-five to forty-five years of age was from eight to ten times the corresponding death-rate among coal and ironstone miners and coincided with the general introduction of machine-drills into Cornish mines. It was well after the turn of the century before wet drilling, fan ventilation and delay in returning to working faces after blasting were introduced in Cornwall and strictly carried out by the cooperation of miners and managers.[20]

In Cornwall after 1949 there were only two producing mines, South Crofty, an amalgamation of several old mines taken over by Siamese Tin

Syndicate in 1967, now called St Piran Mines after the patron saint* of tin miners (see Plate 9). South Crofty for many years operated under a Ministry of Supply contract and Geevor Tin Mines Ltd founded in 1911 to reopen and operate Cornwall's most westerly mine. It produced a thousand tons of tin concentrate in 1969. It is not known precisely when mining started there—the date 1791 is cut in the wall of an old working. In conjunction with Union Corporation it is currently opening up the old Levant mine whose workings extend a mile under the sea, and where efforts were made to seal off the sea water. Some years ago this company, in association with two Malayan tin companies, explored the tin stockwork at Cligga Head on the north coast near Perranporth, which was worked centuries ago when stream tin mining began to decline.

Consolidated Goldfields Ltd, after a long and exhaustive study of a number of prospects, started on 1 October 1971 to produce at Wheal Jane near Truro, the first tin mine to be brought into production for over fifty years (see Plate 10). There a strong lode, well known at adit level for many years, was proved to extend in depth.[21] In times past the complex ore baffled many generations of tin ore dressers and caused it to be abandoned, but because of new improvements in tin metallurgy it is hoped to work it successfully. All the buildings are sited inconspicuously and precautions taken to avoid dust, noise and effluent water.

With tinstone from the stream works little or no dressing and no separation from other metallic minerals were necessary as they had all been oxidized and washed away. It could be sent direct to the smelters after the 'tinners' had cleaned it and concentrated it.

It was quite otherwise with the mine ore from lodes wherein the cassiterite was associated with a variety of other minerals, ores of copper, lead, zinc, tungsten, iron and antimony, also arsenic and sulphur.

The first operation was to crush the ore. Stamps were originally vertical timbers shod with great cast iron feet mounted in batteries of four or more, lifted and dropped by cogs on a water-wheel driven shaft. However, stamps were all metal and driven by steam as early as 1813.[22] Crushing rolls were introduced in 1804 and in 1825 circular buddles of fine mesonery came into use. Rotating trommels† distributed the feed and the concentrates were produced by gravity and water. Large dres-

* A missionary sent by St Patrick to Cornwall in the fifth century A.D. to convert, the heathen Celts. His oratory was near Perranzabuloe.

† A cylindrical revolving sieve for sizing ore.

sing floors with a plentiful supply of water served several mines, and remains of the early tram roads from them, with pairs of stone sleepers for narrow gauge tracks can still be seen.

Horse and muscle power were superseded by steam and much later by electricity, but in more than one large mine the dressing floors were primitive in the extreme. Even in the 1870s tin ore was often sold 'in the stone' to owners of stamps because no dressing equipment was installed at the mine. Many dressing floors were not roofed in and as late as 1892 'bal-maidens' were breaking rock by hand at 4*d.* a ton. Generally however the ore was crushed by stamps.

The dressing floors of the big mines were large and complicated. The ore was selected underground and again on the surface, sometimes on a picking belt. After crushing in stamps and rolls the sized products were concentrated on shaking tables, the middlings treated in buddles, frue vanners, slime tables and rag frames. Middlings from all these were reground in tube mills for reconcentration in water. To remove sulphur and arsenic the concentrates were roasted in calciners, ovens or reverberatory furnaces. Magnetic separators removed wolframite and other minerals of varying degrees of magnetic permeability. Upward current classification was effected by Spitzkasten and slimes treated on round tables. Everything was done to improve the quality of the concentrates to get a better price from the smelters. It has been calculated that in the sixty years preceding 1918 from £20 million to £25 million worth of tin was lost in tailings from the dressing floors. For generations studies by many scientists have been made to improve this situation.[23] In 1903, at the mechanized dressing floors of Dolcoath 526 rag frames were introduced and the manager, Arthur Thomas, claimed highly satisfactory results. Flotation gave discouraging results. Hatfield had some success in dielectric mineral separation.[24] Separating wolframite by caustic soda was not successful and a number of other chemical methods of extraction proved too costly for practical use. A sink and float process using a medium with a specific gravity between cassiterite and the sulphides present has been suggested. In some cases the concentrates were treated with dilute sulphuric acid to improve the grade and get rid of as much of the impurities as possible.

The recovery of fine-grained cassiterite has always been a major, and as yet unsolved, problem in Cornwall and in the heyday of tin production, in the early 1870s, a total of forty-one separate concerns worked the Red River for 'slime tin'.[25] These tin streamers, not to be confused with their former kinsmen working the rich eluvial deposits, were more capable dressers than the miners as it was not uncommon for a 'streamer' to make a handsome profit from treating the waste from a mine making a recurrent loss (see Plate 13).

Since World War I there have been many attempts to recover some of this fine tin washed downstream from earlier times. Between the wars the dredging of Hayle River and, after World War II, the treatment of Gwithian beach sands and the sands of the sea under St Ives Bay were all studied. Assays from systematic sampling gave positive tin values and intensive studies were made on how best to collect and concentrate the tin but none of the many schemes was successful.

The Carnon river rises in Carn Brea near Camborne and took the tailings from a number of mines. It is only ten miles long when it empties into Restronguet Creek, a branch of Falmouth harbour. Here again exhaustive trials failed. Attempts were made with a bucket dredge, a gravel pump on a pontoon and stationary gravel pumps. An effort was started in 1919 and continued intermittently until the beginning of 1924, but with no satisfactory recoveries.[26] However, currently Hydraulic Tin Ltd is producing 100 tons a year of tin concentrates from the Carnon Valley.[27]

The method of selling the tin concentrates from the many mines throughout the boom and for many years afterwards was known as tin ticketing. The mines produced their concentrates for sampling and the samples were taken by the smelters' representatives. The parcels of each mine's concentrates were sold at Tabbs Hotel in the centre of Redruth. The sales used to be held fortnightly. The merchants and agents wrote on a slip of paper or 'ticket' the price they were prepared to pay for a particular parcel the details of which were announced by the chair. The writer of the highest priced ticket was considered to be the buyer. If two bids were the same, which was not uncommon, the parcel was divided between them. The last sale was held on 24 January 1921.

In the beginning, mining and smelting were conducted by the same person or group. The most primitive, Bronze Age method was to dig a trench on a hillside exposed to the prevailing wind, line it with clay, fill it with brushwood and a pile of logs of wood. The brushwood was lit and when the logs were burning fiercely, filling the trench with glowing embers, tinstone was thrown on the fire at intervals. More wood and more tinstone were added until sufficient metal had accumulated. Then the fire was raked out and the molten tin ladled into a hole in the ground or a clay mould to solidify. Later, the shallow hole became deeper to confine the fire and the blast was led through an opening above the base, the molten metal flowing out through a still lower hole.

Later still, a small cylindrical furnace 3 feet broad and 3 feet high, the so-called 'Jews' House,' was found easier to work than just a hole in the ground. The blast was induced by rough bellows worked by hand. Few such furnaces remain intact, but the ruins can be recognized by the remains of the hearth and often by ingot moulds made out of granite blocks and left there as useless when the smelter was abandoned. The blowing houses predominant in the mid-fourteenth century were rude structures of rock and turf with a thatched roof. The furnace inside it was made of massive blocks of stone clamped together with iron and called 'the castle'. The tinstone and charcoal were laid layer upon layer and, after setting the fuel alight, the charcoal was fanned by large bellows 8 feet by 2 feet 6 inches wide worked by a waterwheel. When the heat was raised to the required temperature and reduction to metal well advanced, the furnace was fed with two to three half-shovelfuls of charcoal and four shovelfuls of ore at a time and after twelve hours several tons of ore were reduced to metal. The molten metal was collected in a large moorstone trough, the 'float', and ladled from it into a 3 feet diameter iron pot with a small fire under it to keep the metal molten. Two or three large pieces of charcoal soaked in water were plunged to the bottom by an iron tool. The violent ebullition caused slag to rise to the surface which was skimmed off. The metal was tested and, if bright as silver and uniform, the operation was complete.[28] The metal then was ladled into moulds to form blocks ready for sale.

Naturally, as nearby forests were consumed charcoal became dearer and more difficult to get. Many blowing houses consumed 4,000 packs of charcoal a year, and in the sixteenth century the colliers went as far as Dartmoor forest where they encamped for weeks felling, piling, framing and igniting the pile. The finished charcoal was loaded on to pack horses, each load being about three bushels, and taken to the blowing houses. Later, at the end of the seventeenth century it was imported by sea from as far away as the New Forest. In the accounts of the St Austell blowing house charcoal cost 1*s* 6*d* a peck in 1771.

At a later stage the metal of the first smelting was cast into slabs, weighing three-quarters of a hundredweight. This tin was remelted in a large pot by a gentle fire. To purify it, the workers raised the molten metal to shoulder height in iron ladles before letting it fall back into the pot. They then plunged large blocks of green applewood into the pot causing a fountain 15 feet high of metal particles which, splashing back, appeared to make it boil. After skimming, the refined metal was then ladled out into bevelled brass moulds which, on cooling, were stamped with the mark of the smelting house. In these blowing houses tin seeped through fine crevices in the furnace and fine particles lodged on the thatched roofs which, every seven or eight years, were profitably burnt to recover the trapped metal. Usually the blower owned the works or rented it from the landowner.

Because of the growing scarcity of charcoal, experiments were made to replace it with coal. In 1640 Sir Bevil Greville experimented with coal but his efforts came to nothing. A German, Francis Moult, with a Mr Lydall, set up iron furnaces to smelt tin with coal at Newham near Truro, which were in production from 1705 to 1715. They employed twenty men at 25 shillings a month plus drink money. Only a Mr Heyden, the overseer, got £3. The landing of coals, clay, lime, bricks and sand from the river Fal was done by barrow women.[29]

A German, John Joachim Beecher, was using reverberatory furnaces at Treloweth in 1695. Angarrack smelting works was started in 1704 and this business was transferred to Calenick in 1711, where the smelter remained for many years one of the chief works in Cornwall. However, stream tin, pure tinstone, continued to be blown in the old way. Polwhele, the historian, relates that blowing houses were in operation at Penzance in 1811 and they continued as long as stream tin was produced. The last blowing house only ceased to operate in the mid-nineteenth century at St Austell. Pryce wrote that tin metal from stream tin fetched ten to twelve shillings more per hundredweight than tin from mine ore in 1778. Up to 1837, the smelter was paid by deducting the amount of metal promised to the owner of the black tin at the coinage. Usually this owner sold his tinbill, a form of promissory note, to the smelter who from 1750 onwards took the place of the old tin merchant.

During the nineteenth century many new smelting works were created to deal with the increased output from the underground mines. There was keen competition in cutting off supplies from one to another smelter. Nevertheless the smelting firms were closely interlinked and nearly all were bankers as well. In Truro there were three smelters, one under Williams Harvey of Hayle at Trethellan near the present gasworks, another called Carvedras belonging to the Daubuz family and thirdly a much older one near Tregolls called Truro Smelting-House. Further to the east, the Charlestown Smelting House on the coast was built to prevent St Austell tin from going to Truro. Ostensibly Charlestown belonged to the London firm of Enthoven and Brassy, but the Bolithos of Chyandour had a secretly held half share. West of Truro was the aforementioned Calenick Smelting House at the top of the Calenick estuary, built to replace a much older one at Newham. Here also the Bolithos had a secret interest. The old clock tower of Calenick Works is still to be seen from the Truro to Falmouth road.

The function of the Angarrack smelting works was to intercept tin ore from the Camborne district and prevent it going to Williams Harvey's Mellanear works at Hayle. Treloweth, next door to the Lamb and Flag public house at St Erth, was placed to intercept tin ore from St Ives and Lelant going to the Mellanear works or to Angarrack or Chyandour near Penzance. Later the Trereiffe smelter was built to prevent St Just's tin ore reaching Chyandour. Neighbouring 'pubs' were supposed to assist in this queer but keen competition.

The small smelters seldom got the ore from the large mines which usually sold it at tin ticketing in Redruth or by private treaty. The lesser smelting houses were kept alive by the little mines, and by scores of stamp and stream works which once a month made bargains with their favourite smelting house in the local pub. The bargaining with the smelting works manager often lasted a whole day with frequent refreshment. Most smelters gave their customers a free dinner, a custom continued up to 1912 at the Three Tuns at Chyandour. When the works closed in that year the pub was pulled down also.

The managers of large mines attended the weighing at the smelters. They tipped the man who offloaded the ore 2*s* for each bag and the smelting house manager tipped the wagoners the same amount, an almost oriental exchange of presents. The tin was cast in three hundredweight blocks to prevent theft and each was stamped with the Lamb and Flag or other house marks. After thousands of years tin smelting in Cornwall ceased in August 1931 when the Seleggan works in Redruth shut down.

The reverberatory furnace is in common use today with coal as fuel and reducing agent. A mixture of 10 hundredweight of small coal to 50 hundredweight of ore is a normal charge. With the doors closed the time to complete reduction is about eight hours during which operation several rabblings take place. The metal is then tapped into a brick-lined vessel, allowed to cool, and the slag skimmed off. The metal is then further treated for the elimination of impurities. The draught is induced by a tall chimney. The melting point of tin is generally taken as 232 degrees centigrade. In some of the largest smelting works anthracite was used for from 16 to 20 per cent of the mineral charge. Waste heat recovery plant reduced the coal consumption.

Between the world wars in Britain there were four principal tin smelters. The Penpoll at Bootle, Lancashire, the London Smelting Company and the Penryn near Falmouth (amalgamated before 1925 under the title of Amalgamated Tin Smelters) and the Williams Harvey plant at Bootle, one of the largest tin smelting units in the world. At the two existing British smelters, besides the small output from Cornish mines, practically the whole output of Nigeria and Bolivia is treated as well as concentrates from other sources. The long established Capper Pass works at North Ferriby in Yorkshire treats low grade tin ores and residues from all over the world.

Dennis gives three stages in tin smelting:

1. The production of impure metal and dirty slags.
2. The treatment of slags to recover metal.
3. The refining of impure tin.

The metal from the first smelting is impure and the slag skimmed from the cooled metal in the float contains from 20 to 40 per cent tin and is retreated in similar reverberatory furnaces with a charge of the slag, dross, coal and limestone or scrap iron. Because of the high penetration of molten tin the furnaces are raised from the floor on pillars to facilitate the recovery of metal that has escaped through joints or cracks. The retreated slag produces a low grade metal known as hardhead and containing 80 per cent tin and 20 per cent iron, a nuisance product. Refining of this takes place in small reverberatory furnaces where air or steam is passed through the metal to remove the iron and impurities. The dross is skimmed off for retreatment.

Large hemispherical pot furnaces are used for liquation. Copper can be removed by powdered sulphur, the coppery dross being skimmed off for retreatment. Lead can be removed by passing chlorine through the liquid metal. The refining processes create fume and dust which is collected in a baghouse or a Cottrell plant for briquetting or agglomeration in some other way for retreatment.[30] The incidence and retreatment of many intermediate products containing impurities to extract the maximum amount of tin naturally adds considerably to costs. The reduction of cassiterite is simple, but the refining of the crude metal to remove impurities is complicated, and for many years the details of this purification remained a close secret (see Plate 14). Of course tin can be refined electrolytically where power is cheap. The best British brands are extremely pure. Williams Harvey's No. 1 is 99.86 per cent tin and Penpoll's 99.72 per cent tin.

In the fifteenth century a great deal of tin was used in making brass (bronze) cannon. The chief use was for church bells and for tinning the inside of kitchen utensils. Pewter did not have so much influence as only the rich could afford such tableware.

There is no substitute for tin as a metal, in alloys, or in the ceramic and textile industries, in spite of suggestions that such metals as cadmium or barium could replace tin in solders and bearing metals. In the food preserving industry tinplate still has no serious rival. Since World War II it has been produced electrolytically instead of by the dipping process. Half the world's tin is used in tinplate. Production began in South Wales about 1700 and the industry was in a secure position by 1800 and predominent in 1850, but by 1885 United States output had taken the lead.[31] However, by better financing and technical improvements South Wales maintains its position in tinplate production.

The demand for solders, bearing metals and printing metal does not appear to be diminishing. Printing metal, an alloy of tin, antimony and lead, has the unique property of slight expansion on solidification giving the characteristic sharp character to type. Solder, an alloy of tin and lead, was known to Pliny, the Roman historian, who refers to the joining of metals by it. Tin is used in terne plate (85 per cent lead, 15 per cent tin), in Babbitt and other bearing metals, bronze, bell metal, pewter, collapsible tubes and as salts of tin in the ceramic and textile industries. In the latter it gives the 'frou-frou' to silk garments. Tin for foil contains 5 per cent antimony and for collapsible tubes either pure tin, tin with 5 per cent copper, or an alloy of 85 per cent tin and 15 per cent lead is used.

Stannous chloride is used as a mordant for dying cottons and linen. Stannous chloride or sodium stannate is used for weighting wools but stannic chloride for weighting natural and artificial silks. In the ceramic industry stannous chloride is an opacifying agent for white enamels, glazes and glass.[32] Only in the foil industry has tin been largely replaced by aluminium. Even so, tin foil keeps sweetmeats and cigarettes in far better condition for longer under trying climatic conditions.

The archaeological relics of tin mining in Cornwall are many, the most outstanding being the tall, gaunt engine houses with their chimneys reaching up against the skyline. Many can be seen from the road from Chacewater to Camborne. There is the old clock tower of Calenick smelter still to be seen and a number of more ancient relics of which the oldest is the Celtic fort of Chun Castle near St Just.

CHAPTER FIVE

Copper

After the Romans left Britain the mining of copper ore in the United Kingdom restarted, as far as is known, in the twelfth century but it did not become an important industry until the beginning of the eighteenth century, during which Great Britain contributed about three-quarters of the world's supply of copper, principally from mines in Cornwall, Ireland and Anglesey. Up to about 1856 Great Britain maintained its position as the principal producer of copper in the world, but since that date the output fell steadily because the richer parts of the orebodies in this country were exhausted.

Between the years 1739 and 1865 Cornwall is said to have produced 7,844,305 tons of copper ore with a value of £50,964,388,[1] the chief district being that of Carn Brea. Other important Cornish copper producing districts were Levant, Gwennap, St Austell, Caradon, and in Devon, where Devon Great Consols was at one time the most important copper mine in Europe. In Anglesey the chief mines were the famous Mona and Parys Mines which ceased production many years ago, but from which a small production of precipitated copper is still obtained. About 150 years ago this copper mine was the largest in the world. There is no evidence that any substantial new deposits are likely to be discovered and at present there are no facilities for smelting copper ore in Great Britain.[2]

In nearly all the western counties copper ores have been found and worked with varying success. The ore is found as impregnated country rock, as pipes and other irregularly shaped deposits, or as lodes.

The impregnated sedimentary rocks are most interesting, as the copper ores form a cement enveloping the sandstone grains in which they occur. They are found in beds of sandstone and conglomerate at the base of the Keuper formation of Triassic age. Important quantities of copper ore have been mined in the carboniferous limestone at Ecton,

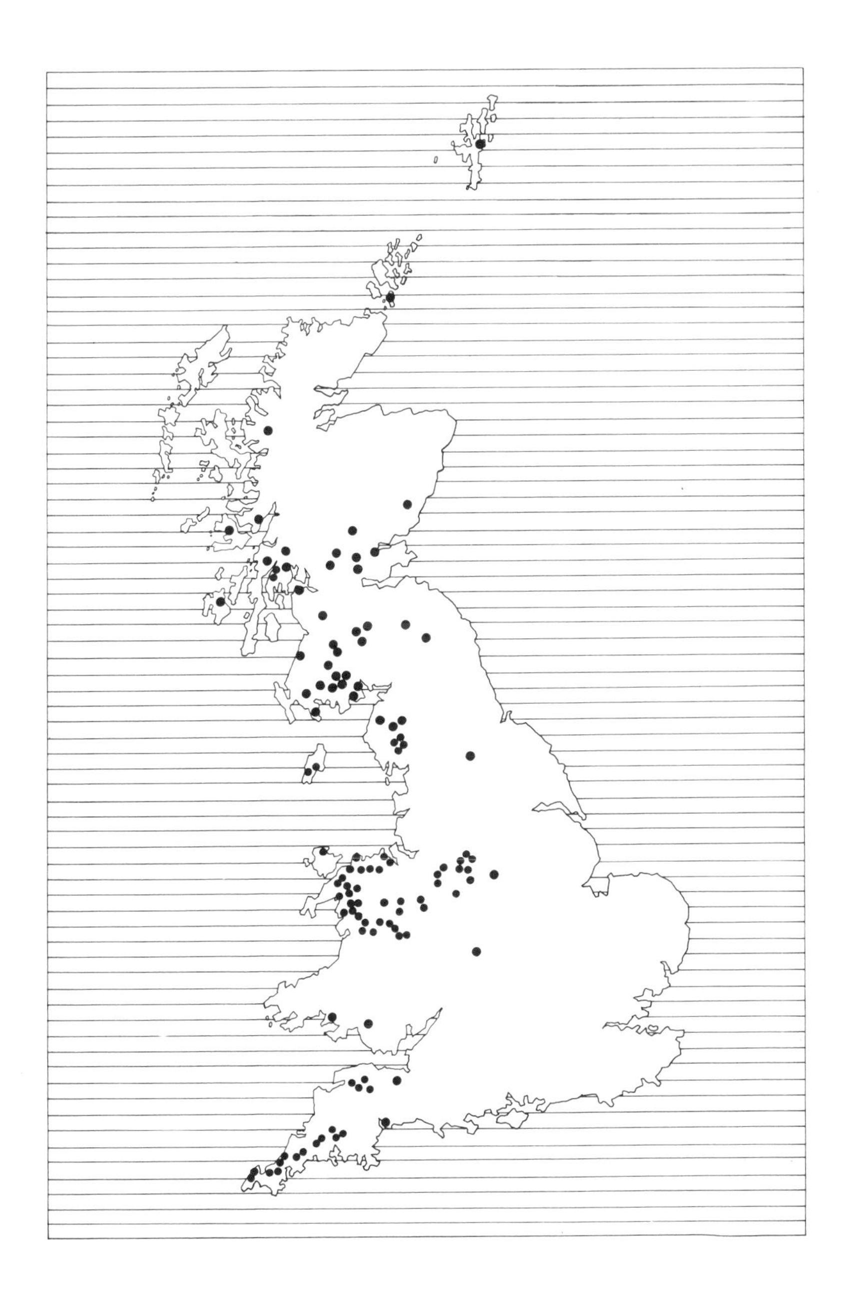

Figure 8 Copper deposits

on the slopes of Great Orme and at Llanymynech, where they formed irregular bell-shaped masses. Fissure lodes were mined in the old Cambrian and Palaeozoic rocks of Wales, the Lake District and the Isle of Man, and in the Amorican rocks of Cornwall and Devon. Evidence exists that copper mines were worked by the Romans at Parys Mountain Llandudno and Llanymynech, whilst tools found at Parys and at Alderley Edge (Cheshire) were considered by archaeologists to date back to the Bronze Age.

In Scotland numerous stone moulds for casting bronze, mostly found in Aberdeenshire, go back nearly four thousand years. At the beginning of the Christian era smelting of native copper ores was general throughout the country, the earliest being in Bute at the Fort of Dunagoil. Towards the end of the sixteenth century the king brought in Frenchmen to increase the output of minerals and in 1583 Eustatius Roche was granted all the mines and minerals in Scotland with the obligation to give one-tenth of all base metals as duty.

In the Shetlands there is a mine called Sandlodge on the east coast of the main island some fourteen miles south of Lerwick. It was opened by Welsh miners at the end of the eighteenth century. They sank a shaft and won about £2,000 worth of copper ore. Owing to the lessee's death the project was abandoned, but by 1800 it was reopened. The principal vein was about 10 feet wide and by 1803 the miners had sunk to over 130 feet. Two cargoes of ore were shipped, the better of which was dressed ore worth £70 per ton. In 1872 a certain John Walker obtained a lease, installed machinery and mined actively for eight years, raising about 70,000 tons of ore. In 1880 £5,814 worth of copper ore was got. The lease was then sold to the Sumburgh Mining Co. which went into liquidation the following year.

In the Orkneys on the south-west corner of the Isle of Burray there are plain indications of copper as reported by George Low in 1774: 'In breaking through a crumbling sort of rock they found pure virgin copper in the form of leaves and sprigs of trees,' a remarkable occurrence.

The second Marquis of Breadalbane believed there was great wealth hidden in the rocks of Breadalbane. His father had closed down the lead mines at Clifton near Tyndrum in about 1798, but the second Marquis had them reopened in 1838 and spent a fortune in keeping them going. Copper had been discovered on the east shore of Loch Tay, and here again he began to exploit the ore at considerable loss. These unprofitable concerns did not dishearten him, and he might be seen anywhere among the hills between Taymouth and Tyndrum with a leathern bag over his shoulder, and a geologist's hammer in his hand, chipping away at fragments of rock, as if his very life depended on it. Numerous trials were made for minerals, extending over many years, at different places on the Breadalbane estates. Chalcopyrite and tetrahedrite were found at Tomnadason* and Corrybuie. These trials all ended in financial failure, although the copper mine at Tomnadason was carried on until the death of the Marquis in 1862. After his death the trustees closed down all mining activities on the estate.[3]

A mine of copper three-quarters of a mile from Bridge of Allan station is the Airmirey mine first worked by the Foullis family in 1661 and intermittently since then. It was reopened by the Caledonian Mining Company in 1805 who produced a quantity of dressed ore smelted at their then new works at Alloa in Clackmannan. In Perth, Stirling and Clackmannanshire there were a number of trials. At Kilsleven on Islay a vein was discovered by a Mr Freebairn, who got 100 lb of good ore in three days, but the mine was abandoned because of the inability to cope with the incoming water. The Murder Lode was discovered by the Omer Copper Company as late as 1911 on a site where the legend was that no grass would grow because of a fight to the death between two women. A rich pocket of ore consisting of equal parts of malachite and sulphides was soon exhausted and the mine abandoned. Ayrshire is pockmarked with small abandoned copper mines, especially near Glengarnock and Kilwinning, but none of them produced a significant tonnage of ore. In Lanarkshire and Dumfriesshire a number of trials are reported, one a mile south of Leadhills, and all abandoned long ago.

There were more copper ores worked in Kirkcudbrightshire than any other Scottish county but the mines were small and few records exist for how long they worked or how much ore was won. Enrick mine was the best of them and is two miles south-east of Gatehouse of Fleet. The lode was discovered accidentally in 1820 and leased to a Welsh company who spent a good deal of money on the mine, raising ore which was shipped to Swansea. A sample from the lowest, No. 5 level, yielded 19 per cent copper. East and west Blackcraig mines are said to have been discovered in 1763 by a soldier while making a military road. Primarily a lead mine, a pocket of chalcopyrite was worked from 1864 to 1865; 28 tons of this ore yielded 3½ tons of copper metal. Heston Island Mine

*Tom-Nan-Daisean: the mound of hay ricks.

was at the north end of a rocky island at the mouth of Auchencairn Bay. An English company worked it in 1845 and sent the ore to Swansea. Williams, in *The Natural History of the Mineral Kingdom* (1810), mentions twenty veins of copper ore in Colvend parish. They were situated at the head of a rocky inlet near Bells Isle and a considerable amount of ore was raised about 1770.

In the Lake District of England the copper ore usually occurs in veins trending east and west in country composed of the tuffs of the Borrowdale series of rocks. The Coniston Mines are the oldest in this district, worked by the Romans and probably by the Ancient Britons before them. They had been long abandoned when Daniel Houghstetter, Queen Elizabeth I's mining expert, came to England. He is claimed to have said that if he had known the richness of the Coniston mines as compared with Goldscope and Caldbeck he would have erected his smelting works at Coniston instead of at Keswick.

He described many works unrecognizable now. God's Blessing or Thurdle Head is said to have been extremely rich, and seven workings all close together at Levers Water, were worked to a depth of up to 12 fathoms. One vein, sixteen inches thick, was largely of malachite. Houghstetter wanted to drain Levers Water to get more ore but this project was never carried out. He was an able mining engineer who set about establishing a copper mining and smelting industry without any local facilities or proper roads. Because local miners were few, he persuaded many German miners from the Tyrol to join him. Though there was friction at first with locals, who thought these 'Dutchmen', as they were called, would keep all the jobs for themselves, soon there was work for all and the 'Dutchmen' married local girls. It was easier to carry goods on packhorses than in wheeled vehicles on the existing dirt tracks. Only the heaviest machinery parts were carted. The Keswick smelter was built at Brigham by the Greta river and so had a plentiful supply of water. A cluster of furnaces, a refinery and eventually a rolling mill were erected on the latest German design. Grassmere mine was opened up and here also a dressing plant was built. Ore raised at the Coniston mines was sent by packhorses to the Keswick smelter.

The mining ground of Coniston, which has produced over three quarters of the total copper ore production of the Lake District, lies approximately within a parallelogram formed by Levers Water, Levers Water Beck, Erin Crag and Red Dell Beck, and is about half a mile long and a quarter of a mile wide. The Bonsor vein produced about half of the total output. This vein was stoped out almost completely over a length of a quarter of a mile and was reckoned to carry a rib of solid chalcopyrite (sulphide of copper) about 8 inches wide. The deepest workings were some 1,250 feet below adit. By 1895 the copper price had dropped below the economic limit for such deep workings, and the ever-increasing amount of magnetite (magnetic oxide of iron) in the ore made gravity separation difficult. So pumping was stopped, the mine allowed to flood, and as the water rose worthwhile pillars of ore were robbed. The water reached the horse level by 1900.

At the Paddy End Mine, said to have richer ore than Bonsor, all available pillars of ore were recovered in the late nineteenth century and by 1908 the workings were in a dangerous state, but good ore was opened up just before working ceased. Triddle Mine was busy in the eighteenth century with a stamp mill by the Red Dell Beck but no work has been done there since 1895. God's Blessing or God's Gift Mine is thought to pre-date Elizabethan times. The mine was rich in silver with a little gold, and it is said that Elizabeth brought Daniel Houghstetter to England in 1564 to improve the working of this Mine Royal. The vein is narrow but rich. In 1907 a few tons of ore were raised when the old workings were examined. Greenburn mine, often called Great Coniston mine, was opened about 1855, worked down to 720 feet and abandoned when the fall in the price of copper made it uneconomic. However, its Sump vein is said to have yielded about £25,000 of copper pyrites.

The mines of the Caldbeck Fells were only small producers of copper ore in the upper part of the lodes. Many have been worked intermittently since the days of Elizabeth I. Roughtongill Mine, lying some 3½ miles south-west of Caldbeck, is thought to have been started as early as the twelfth century. It was idle when the Germans took it over in 1566, after which it was worked with vigour for a number of years. Little is recorded until 1710 when Lord Wharton worked the Silvergill vein successfully. From 1873 to 1877 the output of copper ores was 1,498 tons.

The copper mines of the Keswick area include Dale Head and the Goldscope mines, both of which were worked by Houghstetter and, in the seventeenth century, for the Duke of Somerset. At Dale Head work was resumed several times in the nineteenth century and it is said that good quantities of ore were raised but money was lost in trying to smelt the ore locally. The ancients seem to have mined all the good copper ore which was above the horizon of the new level driven by a Mr Johns in 1919. Queen Elizabeth claimed it as a Mine Royal as mentioned in Chapter 2. Smelting was carried out locally. The mine was operated by different parties until the Civil War of 1650, when the Roundheads are said to have destroyed the smelter. In 1690 William of Orange brought over Dutch miners who are said to have remained until the mine was no longer worth working.

Alderley Edge copper mines, exploited in the middle Bronze Age, are the oldest copper mines in Britain (see Plate 15). Stone hammers and axes were found; the hammers weighed from one to fifteen pounds, though usually about four or five pounds, and were grooved for hafting. A roughly used oak shovel was also found. There was no trace of metal tools, the stone hammers being good enough to break up the soft sandstone impregnated with copper carbonate. The stone tools were coated with a film of green copper carbonate and old timbers were similarly encrusted. Opinions differ as to whether these deposits were worked in Roman times. There are three separate mines, West Mine, Wood Mine and Engine Vein Mine. The first workings to be dated are early eighteenth century when a tunnel was cut from Dockens Wood to the engine shaft near the Great Quarry and although there are no records earlier than 1813, Henry Holland mentions adits driven through the hill a century previous to that date. Most of the stopping both in West Mine and Wood Mine was done between 1857 and 1877 and little mining has been done since.

In Wood Mine the series of beds are ten with sandstones and conglomerates alternating. The top five are rich in copper ores and cover some 30 feet in thickness, then two feet of red clay overlies another 10 feet of less important impregnated sandstone and conglomerate. The mineralization is patchy. Galena and lead carbonate are occasionally found in sufficient quantities to be worth extracting.

At the entrance of West Mine all the lower beds of sandstone have been excavated and a huge cave-like hollow extends back into the mine for many yards. From the entrance to the boundary across the strike the distance is over 1,300 feet. So much stopping has been done that only about a quarter of the ground remains, mainly as pillars supporting the roof.

Wood Mine is a quarter of a mile to the east but not connected, the adit extending to 4,500 feet. The haulage way runs in a cutting from near the mills to Windmill Wood where it enters the tunnel. The rock was dug with pick and shovel and although often friable the walls stand without timbering and the roof is strong enough to stand unsupported.

The richness of the ore varied considerably in both sandstone and conglomerate, the workable rock averaged about $1\frac{1}{2}$ per cent of copper metal, but locally enriched samples have yielded 7.5 per cent. The assemblage of minerals points to the conversion of the sulphides of the metals by atmospheric waters and the resultant minerals cement the sandstone and conglomerate into hard rock.

The broken rock was ground and screened before being macerated in a solution of hydrochloric acid, then filtered and the sap-green copper liquor pumped into wooden reservoirs. Scrap iron was thrown into these, causing the copper to precipitate in metallic form, which consisted of 80 per cent copper and 20 per cent iron. The recovery was about 75 per cent of the copper in the ore.

The total output of copper ore at Alderley Edge from 1857 to 1877 was 168,269 tons and the copper obtained from smelting over these twenty years was 2.1 per cent of the ore mined or some 3,500 tons of metal. The plant consisted of jaw crushers, screens, acid tanks and crystallizing tanks. Mining was temporarily renewed in 1915 with an output of 35 tons of ore. The presence of arsenic in the ore was a great drawback to its treatment.

Mottram St Andrew is two miles east of Alderley Edge. The exposure in a quarry there is 6 feet thick and is between two bands of marl. Ores of copper, cobalt and lead were got from the conglomerate at the base of the Keuper series. The ore was richer than that at Alderley Edge, being said to contain up to 22 per cent copper with an average grade of 5 per cent. The copper was recovered by a treatment similar to that at Alderley Edge.

In Shropshire at Grinshill Copper mines, 160 yards SSW of Clive Church, the copper ores were found in grey sandstone at the base of the Keuper division of the Trias. Most of the copper-bearing rock was confined to a north-west fault and consisted of millet-seed quartz grains cemented by barytes and carbonate of copper. A former mine captain said the ore often swelled out into masses but the average stopping width was 20 feet. The mines were worked for several centuries and in old levels seventeenth century tobacco pipes and old mining tools proved that a great deal of ore had been removed. The ore was crushed, screen-ed and carried to two rows of twelve tanks, made of Yorkshire flag-stones with false bottoms, where it was treated with hydrochloric acid. The solution then went to wooden troughs where the copper was precipitated with scrap iron. The precipitate was dried in kilns, packed in barrels and sent to smelters.

There is a mine 5½ miles south-east of Oswestry called Eardiston. Again the country rock belongs to the basement beds of the Keuper series of Triassic age. Some malachite occurs, but a thin steeply dipping blue clay layer cut out the ore about 100 feet below surface. From 1841 to 1845 it produced 2,500 tons of ore and intermittently until 1865. When work stopped it was producing 30 to 60 tons a month. The analysis of an average sample gave 11.5 per cent copper.

The most important mine where the ore was in irregular masses or pipes is in Staffordshire on Ecton Hill near the boundary with Derbyshire, a famous and historic copper mine. The country rock is carboniferrous limestone. The hill is an anticline whose axis runs north and south and the beds are bent into many sharp folds, sometimes as sharp as the letter V. Prince Rupert, nephew of Charles I and a man of many parts, introduced gunpowder to metal mining in Britain at this mine about 1638, having sent for German miners to work it. It was reworked about 1707 but abandoned. In 1739 it was rediscovered by a Cornish miner who soon gave it up, but a second set of Cornishmen were much more fortunate. They sank a shaft 600 feet deep and drove an adit to find great quantities of copper ore which increased as they sank deeper. The copper ore did not form regular veins but as it went down, widened in the form of a bell. In 1769 the mine was entered by a nearly straight adit about 1,200 feet long, which led to a vast cavern 750 feet high and 480 feet below the level. The pipe extends 900 feet below the river or 1,350 feet below the surface at Clayton's Shaft. In the old Ecton or Dutchman's mine three great open chambers exist formed by the removal of copper ore. Evidence of a zonal distribution of the ores of copper, zinc and lead are remarkable and form a parallel with some Cornish mines. Some secondary enrichment appears probable because from 1776 to 1817 the ore was said to be 15 per cent copper, but in the deeper levels it was of low grade. Some output figures are from 1760 to 1768 5,862 tons copper ore and from 1776 to 1817 53,875 tons copper ore.

In Shropshire is a mine famous for its links with Roman times. At Llanymynech mine there are old shafts and opencast workings on the hill of that name about a mile NNW of Llanymynech Church. The country rock is carboniferrous limestone. In a great artificial cave, with galleries meandering from it in search of ore, Roman coins were discovered. Near the north-west part of the hillside are numerous large pits in the shape of inverted cones, supposed to be Roman workings, when they followed the vein of copper and a level driven to the old opencast workings above Offa's Dyke. A local miner described the Roman underground workings as a series of galleries opening into great chambers. Relics found at this ancient mine suggest that the miners were confined underground and that they were sent there as punishment for political or criminal deeds – slave labour!

Many mines were worked for copper in the veins and lodes of Wales and its borders. The most important are the Mona and Parys mines of Anglesey. The Romans worked these mines and built a fort at Segontium (Caernarvon) specially to protect them. They were almost certainly worked by the Ancient Britons before the Roman conquest during the Bronze Age. Bun ingots of copper of Roman age have been found at Trysglwyn farm. It is thought the Romans worked it as a village industry, the natives mining and smelting the ore, which was then collected as tribute by the authorities. There are two opencast pits, among the largest in Europe, known locally as the 'Great' and the 'Hillside'. Their origin is due to a great collapse of roof caused by the miners robbing the supporting pillars on being refused fresh leases. The mine commenced by the sinking of many shafts and the driving of numerous levels. Only after the great collapse was it worked opencast (see Plate 16).[4]

No work seems to have been done between Roman times and the eighteenth century when prospecting started in 1757. The early ventures did not meet with much success until the 'Great Lode' was struck in March 1768, when prosperity began. In a few years Parys mine became the most productive in Europe, producing some 3,000 tons of copper metal a year. From 1770 to 1790 Anglesey, and not Cornwall, dominated the world's copper production, all the more menacing to the disunited Cornish mines adventurers because the Anglesey mines came to be controlled by one man, the astute and capable Thomas Williams. Although the ore was not rich, this was more than offset by the occurrence of two vast masses of ore less than 7 feet below the surface so that it could be quarried rather than mined and little exploration or drainage was necessary. Moreover the deepwater port of Amlwch was only a mile and a half distant and the coalfields and smelting potentialities lay not sixty miles away across Liverpool Bay. Williams, realizing the low price the Swansea smelters paid for their ore, set up his own smelting works at Amlwch and elsewhere to wrest the copper market from the Swansea ring. To fight him the latter forced down the price of Cornish ore and the fiercer the battle grew the more poor Cornwall suffered. Williams claimed he could flourish with copper at £50 a ton. At this time the output of the two Anglesey mines of Parys and Mona was little less than all of Cornwall's combined and with his interest in smelting at Amlwch, Liverpool, St Helen's and Swansea, Williams held half the copper trade of Britain.

There were twelve lodes and the deepest workings were those on the "Gwen" shaft going down to 570 feet below sea level. The collar of the Cairns shaft on the summit of the hill was 480 feet above the sea, so the vertical extent of the mine was over a thousand feet. Horizontally, it was nearly one and a half miles west to east and a third of a mile north to south. The western workings were the Parys mine, the eastern Mona mine. The southern and western group of veins were largely bluestone and the northern and eastern, quartz, pyrite and chalcopyrite impregnations. The total contents of the pits cannot have been less than 90 million cubic feet. The whole mine appears to have been exhausted about the 1880s. The general grade of ore was only from 3 to 5 per cent copper. An area of over 17 acres had been quarried to an average depth of 120 feet yielding 2½ million tons of ore and a vast profit of over £2¼ million. The North Discovery Lode was by far the richest, having as much as 20 to 25 per cent copper. It was more like a true fissure vein than any of the other lodes. Secondary enrichment was not observed and the ores were regarded as primary deposits.

The only copper now won in the area is precipitated from copper-sulphate-bearing solutions drained out of the mine.[5] The out-flow of this is controlled by a dam, the solution flowing through a launder half a mile long to precipitation tanks made of brick and cement with wood partitions. Scrap iron is placed in the tanks and the copper precipitated. The effluent is pumped to a series of ochre ponds. The precipitated copper is collected thrice a year from the top tanks and yearly from the lower tanks. The copper precipitate is dried in kilns, and averages 60 per cent copper, the rest iron.

At Rhos Mynach mine two and a half miles ESE of Amlwch, there are three lodes and the workings show that the district has been subject to severe and repeated crushing. Native copper suggests that ultra-basic rocks played a role in the emanation of the orebodies. This mode of occurrence is familiar. It is seen at the Lizard and at Lake Superior on a grand scale. Samples gave 5.6 per cent and 2.6 per cent copper. A picked sample from the middle lode gave 11.37 per cent copper. There are three shafts and an adit that opens near the sea. The output during 1917 was about 80 tons.

In the county of Merioneth there were a number of gold mines which started as copper mines, of which a number are mentioned in the earlier chapter on gold. Gamallt mine is four miles from Ffestiniog. There are two adits and the lode contains mixed sulphide ores, galena, blende and copper pyrites which is abundant and is said to have yielded 17 dwt of gold a ton of ore. Cwm Cynfal mine is about three miles east of Ffestiniog station—at one locality here a big pocket of ore was discovered and worked out. The ore was chalcopyrite.

Dol-Frwynog mine is two miles NNW of Llanfachreth Church on the west bank of the Afon-Wen where four levels have been driven into the mountainside above the river. There was a 4-foot-wide main lode and the longest crosscut cut five parallel veins. Gold has been associated here with chalcopyrite and pyrites.

The Turf Copper Works is two miles NNW of Llanfachreth Church at Dol-Frwynog. A remarkable mode of occurrence was the metallic copper at the Turf mine, where leaves, wood and nuts were preserved by the metal replacing the vegetable substances, while much of the peat was so richly impregnated with the carbonates of copper that it was cut into blocks and sent straight to the smelters. A peat bog forms most of the bottom of the valley. Normally the turf was pared from the surface and burnt in kilns and being partly saturated with a copper compound, a residue with a valuable copper content was left in the ashes. Many thousands of pounds' worth were extracted. The search for the source of the copper in the neighbouring hills failed and probably does not exist. The water percolating through the rocks and rising in springs carried the sulphate of copper in solution derived from the sulphide which was diffused through the mass of the surrounding hills. In one year 2,000 tons of ashes were sold at a profit of about £20,000. In Henwood's time, *c.* 1870, turf that did not yield more than 2.5 per cent copper in the ashes was considered too poor to be wrought and at that time an enormous quantity was left untouched.

Glasdir (Green Land) mine, a mile west of Llanfachreth Church, is an impregnation of the country rock of Lower Silurian age. The ore is about 1 to 2 per cent copper with malachite which stains the water issuing from the adit a bright green colour. A little gold occurs with the copper. Two opencasts about 200 feet deep were drained by an adit. The bottom of the shaft was 620 feet below adit and the mine was worked by levels and stopes to 500 feet below adit. The ore was treated at the mine by an Elmore flotation plant and some concentrate smelted on the spot. The orebody was bowl-shaped over 500 feet long by 5 feet wide at a depth of 500 feet. From 1872 to 1914 the copper ore output was nearly 10,000 tons. In 1914, the year the mine closed down, 1,600 tons of copper ore yielded 900 ounces of silver and 900 ounces of gold.[6]

In Figre (Vigra) mine there are two intersecting quartz lodes, one almost due east and west,.the other can be traced for 1,000 yards on the east into the continuation of St David's lode. The ore is mainly copper pyrites and contains gold. The lode is a clear cut fissure needing little support. Panorama mine is about two miles east of Barmouth station. The gold lode was in places well mineralized with galena, blende and chalcopyrite and traceable for over half a mile. The north copper lode parallels the gold lode, traceable for a mile and contains chalcopyrite and pyrrhotite. A north-south lode intersects the copper lode and to the west of the junction the main lode becomes richer in copper.

The Drws-y-Coed mine in Caernarvonshire was five miles west of Pen-y-Groes station. The principal ore was copper pyrites and the concentrates contained up to 10 per cent copper. The mine was de-watered by a Cornish pump. Between 1855 and 1909 the copper ore produced was 8,696 tons. Taly-Sarn mine, about two miles east of Nantile, was, with the Drws-y-Coed mine, worked by the Mining Corporation of Great Britain. The ore was chalcopyrite, richer than that of the former mine, and treated at Drws-y-Coed.[7]

Llandudno copper mine is a mile from Llandudno station on St Orme's Head. The country rock is carboniferrous limestone. The copper ores were confined to beds of crystalline limestone dying out when the fissures traversed non-crystalline rock. Thus there were successions of productive and barren filling and, when crossveins intersected the lodes, productive material extended to the whole of the group constituting a large deposit of copper pyrites. This was a Roman mine and from evidence found in cavernous openings it is believed to have been worked by slaves and convicts who were confined in the mine. This mine was profitable for many years until unexpectedly the sea broke into the workings along hidden fissures and the mine had to be abandoned. According to *Mineral Statistics* this mine produced 13,558 tons of copper concentrates between 1855 and 1869.

In Denbighshire several copper-bearing veins occur in the Llansannan and Llanfair-Talhaiarn area some eight miles west of Denbigh. At the old Llanfair Talhaiarn lead and zinc mine, copper ore was found on the 85 fathom level. In neighbouring Flintshire also small quantities of copper ore were found. An extensive deposit was worked at Graig-Fawr near Talagoch. In Cardiganshire and West Montgomeryshire copper ores have been raised at several mines, principally at South Daren and Esgairfraith, but the total output from 1857 to 1913 was only 6,055 tons. At Esgairfraith the average grade of ore was 11.9 per cent copper. At South Daren the grade was 9.4 per cent.

On the Isle of Man the copper mines were unimportant. The maximum output appears to have been from Great Laxey mine in 1865 when 1,317 tons of copper ore were reported. The copper ores occur either as thin stringers or finely disseminated on the outskirts of the lead orebody. At Laxey it heralded the impoverishment of the lode. The copper pyrites was mainly obtained south of the engine shaft. From 1845 to 1888 the output of copper ore was 8,864 tons. Previous to that Laxey sold 3,412 tons at Swansea between 1822 and 1844. The Bradda mine is on Bradda Head north of Port Erin Bay, where copper ore only occurred in spots. A little native copper was found in part of the highest point above sea level and worked intermittently. From 1850 to 1883, 1,417 tons of copper ore was sold.

On Exmoor the Saxons certainly mined copper at South Molton at a time when there was a considerable population at North Molton. The Tabor Hill adits were made by Germans in the time of Elizabeth I when the Bampflyde and Poltimore mines produced copper. The North Molton copper mine was worked with good success for many years and only closed in 1820.

A mine called Wheal Eliza in the Barle valley below Simonsbath in Exmoor worked in the early 1880s when a little copper was raised. This mine was once mooted as a gold mine and in 1854 the *North Devon Journal* reported that an important lode had been discovered in the Wheal Eliza copper mine on Exmoor. But it was never really a successful mine. The Brimley mine was more important and raised 13,445 tons of copper ore, valued at £13,618 betweeen 1826 and 1887, the ore being sent to South Wales when prices were good. This mine closed in 1893.

Derbyshire miners are reported to have come to the Quantocks about 1670 in early attempts to mine copper ores in what was then a purely agricultural area, a starving countryside, where farm labourers and their families existed on 8*s* a week. Surface veins and the backs of lodes showed green stains across the dirt roads and highways where the bedrock reached surface. In 1719 the trustees of Alec Luttrell granted a twelve-year lease of the 'Copper Ore Pitts and Mines found at Perry Hill at Quantozhead'. This is the only evidence that such a mine existed because at Perry Hill there is no sign of workings.

In 1725 a foreign traveller, Henrik Kakzmeter, recorded that not far from Bridgwater in Somerset at a place called Stowey there was a copper mine. At the village of Nether Stowey there was a ruined engine house and opposite a dwelling that was the Counting House (i.e. offices) of the Buckingham copper mine. The records of this Dodington copper mine extend from 1786 to 1810 and from 1816 to 1825. The mining was financed by Richard, Marquis of Buckingham, who inherited the Dodington estates and was interested in mining ventures in Cornwall. The quaker Fox family, who also had interests in Cornish mines, had a large holding in the venture. There was interminable correspondence between the Marquis and his agent, Camplin, about the Cornish mine captains who, not unnaturally, tended to exaggerate the value of the ore in the mine. In 1786 the mine was drowned but reopened and worked till 1810. There was an ill-fated revival from 1817 to 1822. The mine closed in March 1821, having cost the adventurers £20,000 of which only £2,500 was recovered by the sale of copper ore.

The true beginning of copper mining in Cornwall towards the end of the sixteenth century coincided with the expansion of the German metal industry which was highly organized in contrast to the primitive state of the brass and copper industries in Great Britain. Sir William Cecil, secretary of state to Elizabeth I, realized this and brought in German miners specially to develop the mining and smelting of copper and lead.

In 1584 a German master miner Ulrich Frose was the manager of a venture at Perran Sands, now Perranporth, mining copper and attempting to smelt the ore there to reduce its bulk for shipment to Neath, where he moved later to take charge of the smelter. About 1580 the Mines Royal leased their mining rights in Devon and Cornwall to Thomas Smith, then Collector of Customs at the Port of London, for £300, and the small smelting works mentioned above was set up at Neath under German management to refine copper produced by the Cornish mines. The biggest problem seems to have been to get enough ore to keep the furnaces working because Frose at Neath in 1585 complained that the works were at a standstill for lack of ore.

During the seventeenth century little is known of the state or the progress of copper mining in Great Britain but it seems to have been moribund as Swedish copper from the famous Falun mine dominated the European market.

Worked on a small and almost secretive scale up to the end of the sixteenth century, copper ore in Cornwall seems to have been regarded as of little value and it was not until the last decade of the seventeenth century that its worth began to be realized. A writer of 1697 was informed that at Trevascus in Cornwall there was more than a thousand tons of ore in a spot 32 feet wide 8 feet deep and of unknown length. A leading character to develop copper mining in Cornwall was John Coster, the son of a Forest of Dean ironmaster, who set up a copper smelting works at Redbrook on Wye, where smelting of both lead and copper with coal was carried out with commercial success by 1687. Small cargoes of Cornish copper ore were soon being shipped to Chepstow. Coster's success was noted by some London merchants who set up the rival English Copper Company nearby about 1691; later it moved to Swansea and became the largest smelting company in Britain throughout the following century. To secure a steady supply of ore Coster took a lease on the Chasewater mines soon after 1700 and he and his son played a large part in developing copper mining in Cornwall. He has been called the 'father' of Cornish copper mining with some justification. He improved the dressing and assaying of the ore and not only extended the use of drainage adits but in 1714 patented with his son an improved waterwheel which aided materially the dewatering of the mines.

As the mines grew deeper the problem of keeping them clear of water became more acute. Waterwheels drove pumps but there was never sufficient river water in the summer to keep them fully at work. So the mines had to rely on the adit and the horse whim to keep them dry.

Thus the introduction of steam-driven pumping engines inevitably followed. The first in the field was Thomas Savery of Devon, who patented a not very successful machine in 1698. More successful was the improved engine built by Thomas Newcomen, a blacksmith from Dartmouth. Newcomen, with the assistance of John Calley, radically altered Savery's design, transforming it into a true steam engine of the atmospheric type. Newcomen was unjustly debarred from patenting his machine, because Savery's patent extended from 1698 for thirty-five years, covering all forms of 'raising water by the force of fire'. So Newcomen formed a partnership with Savery. His first engine to be erected in Cornwall was in 1720 at Wheal Fortune.* It had a 47-inch diameter cylinder and, at fifteen strokes a minute, pumped from thirty fathoms. Such crude machines used steam at very low pressure and were safe. However, they consumed enormous quantities of coal owing to the inherent thermal inefficiency of the design. A large engine would consume 12 tons of coal a day. During the introduction of these engines, so costly to operate, many adventurers turned afresh to adit drainage. The County Adit was started in 1748 by one of the Williams family at Bissoe Bridge on a branch of Falmouth Harbour and was gradually extended to the west boundary of Poldice. During the next half century it was driven intermittently until it drained forty-six mines and reached a total length of thirty miles, the farthest mine being five and a half miles from the mouth at Bissoe. By benefiting so many mines the sharing of maintenance costs became extremely complicated.

Large copper deposits were found outside what had hitherto been considered the mineralized zone. Pool Adit, later Trevenson mine, was to prove extremely rich about 1740. It helped to establish the wealth of the Basset family who held all the shares. They gained a regular income of over £10,000 a year from this mine and an additional £4,000 a year from Adit Dish, a great fortune in those days. Some adventurers actually drove adits, not to drain a mine but to seek fresh deposits. Driving an adit was slow but cheap, as some £300 a year was scarcely felt if spread amongst a dozen adventurers with the chance of amazing profit if they struck a rich deposit of copper ore. The average size of an adit was 6 feet by 2 feet 6 inches, and, with only a single man able to work at the face, took years to complete. The driving of Penventon adit took thirty years but cost only £165 18*s* 6*d* for 326 fathoms length, plus ten ventilation shafts at £3 each.

The copper deposits discovered on Parys mountain in Anglesey in 1768 were so easily exploited that overnight the Cornish mines were put into second place and this continued for some twenty years. The only hope for Cornwall was to reduce costs and the obvious cut was in the coal bills. So Cornish engineers concentrated their efforts on improving the Newcomen engine. Hitherto, to cope with the increase in depth of the mines they had increased the size of the pumping engines and thus

* Although Joseph Carne in his early history of Cornish copper mining states that an earlier Newcomen engine worked at Huel Vor tin mine in Breage between 1710 and 1714, this has never been confirmed.

the coal bills. The elder Trevithick, Budge, the elder Hornblower and Nancarrow all turned their minds to getting the greatest amount of steam from a bushel of coal. Smeaton by 1775 had so improved the Newcomen engine that it was almost twice as efficient. His 72 inch engine at Chasewater pumped the mine down to 60 fathoms (360 ft), a good performance for this type of machine. Many of the older mines could not be worked deeper until there was a more powerful and economical means of dewatering.

This was the position when the Boulton and Watt partnership appeared upon the scene. The first approach was from the Cornish adventurers who, in 1776, sent a deputation from Cornwall to visit Boulton's works at Soho near Birmingham. As a result, the first two engines were ordered, a 52-inch engine for Ting-Tang mine near Redruth and a smaller 30-inch engine for Wheal Busy. Owing to the difficulty of transporting the 52-inch cylinder, the smaller engine, though ordered second, was completed first. It was set to work in September 1777, thus becoming the first Boulton and Watt engine to operate in Cornwall. The whole Cornish mining fraternity closely watched the progress of these two engines. Both were successful and greatly reduced coal consumption. Watt's success naturally had a mixed reception because the livelihood of many Cornish engineers depended on erecting and maintaining the Newcomen engines and they were loath to acknowledge that for equal size the Watt engine was three times more powerful. The most notable order was for five large engines for Consolidated Mines (Wheal Virgin) which were all working by 1782. By the end of 1783, twenty Boulton and Watt engines were at work so successfully that out of seventy-five Newcomen engines, only one survived at Polgooth and it is doubtful whether that was still at work.

Altogether, Boulton and Watt erected fifty-five engines in Cornwall, of which the great majority were single-acting pumping engines.

These new engines were bought outright by the adventurers, but in addition Boulton and Watt claimed a sum of money representing a third of the value of fuel saved. They thus aimed to reap a good harvest from their patent. Consolidated Mines agreed to pay £2,500 a year for the use of the five new engines mentioned above; a big sum but only a fraction of the £11,000 the mine owners saved on coal costs. William Murdoch came to work for Boulton and Watt in 1779 and supervised the erection of many of their engines. He lived in Redruth, the then centre of the copper mines. He was always popular at the mines and was sent for if anything went wrong with an engine. If Watt went instead the miners were bitterly disappointed. Matthew Boulton and James Watt not unnaturally joined the ranks of the adventurers, but Watt, a dour and dyspeptic Scot, had no love for the Cornish and they disliked him. Boulton, the Birmingham businessman, thought the administration of the Cornish mines poor and wanted to improve their accounting and commercial management; he particularly disliked the merchant adventurers selling the miners goods at an inflated price, a practice he had never permitted at his Soho works.

In the 1790s the output of Anglesey mines began to fail and between 1785 and 1800 the price of copper had doubled. But the copper mine adventurers were still saddled with Boulton and Watt's monopoly and now looked at the dues on fuel saving as an iniquitous tax. Their energy was now bent on ridding themselves of this Midland incubus. Watt's patent expired in 1800 by which time the firm had entirely lost the goodwill of the clannish Cornish mining community. Nothing remained for the partners but to collect the long delayed dues owed by a number of mines, a task which was ably accomplished by their respective sons.

An odd character was Rudolf Erich Raspe who robbed his master in Central Europe of 2,000 dollars. Arrested, he escaped to England where he became assay master at Dolcoath in 1782 and, in 1791, published a translation of Born's work 'A new process of amalgamation'. Moving to Scotland he persuaded Sir John Sinclair to sink a lot of money in a mine he had 'salted'. Detected, he fled to Donegal and died there in 1799. Sir Walter Scott's novel *The Antiquary* is based on the Sinclair-Raspe episode.

By the end of the century Watt's engine was wellnigh supreme although many improvements had been introduced by Trevithick and other Cornish engineers. A Cornishman, Arthur Woolf, experimented with compound expansion and introduced a new high standard in engine construction. He joined John Harvey's growing engine foundry at Hayle as superintendent. Several other engineers, trained by Woolf, contributed to the technological advances in the early decades of the nineteenth century, such as the introduction of small, shallower fireplaces with much greater draught — needing taller chimney stacks. Grose, one of Woolf's trainees, introduced the lagging of cylinder, steampipes and boiler to conserve heat. So by 1820 the Cornish beam engine—basically designed by Newcomen, developed by Watt, perfected by Trevithick and further improved to work on high pressure steam by Woolf—became world famous for its simplicity, economy and efficiency. The major problem of dewatering mines was mastered. In 1818 Dolcoath was mining at a depth of 227 fathoms, not much less than 1,400 feet. An abandoned mine allowed to fill with water requires time and money to be rehabilitated. North Down, not a deep mine but extensive, cost nearly £20,000 to put in working order again after only a brief closure.

The great problem of keeping the workings dry having been overcome, the other serious problem was the transport of coal to, and ore from, the mines to the coast for shipment to Wales. At the turn of the century Boulton estimated that 2,500 horses and mules were employed by the mines. Churned up by the hooves of the heavily laden animals, the tracks were rendered impassable by the winter rains and the expenses and delays of such a system were detrimental to the rapid growth of ore output. The great increase in the use of tramroads in South Wales indicated the solution. Packmule transport had become so costly that one of the smelters refused to buy any more ore from the rich but isolated Tresavean mine.

In 1806 an iron tramroad from Dolcoath to Portreath harbour was proposed, actually surveyed, but never built. Three years later another tramway was built from Portreath's small, and in winter, stormbound, harbour to link up with the mines of Scorrier and St Day, extending to Wheals Unity and Poldice, then in their prime. This pioneer tramway was worked by horses throughout its life. It was owned by the Portreath company and was leased to the Foxes of Falmouth by the Bassets who had built the harbour several decades previously. It paid handsomely from the start as the animals pulled many times the weight of about 300 lb which was all that they could carry.

In 1824 John Taylor, the London adventurer who had taken up the old Consolidated Mines and several others nearby five years before, built a tramway from Redruth past his mines to Devoran on a tidal inlet of the Fal estuary. In its first year this 4 foot gauge road carried over 50,000 tons of ore and 20,000 tons of coal. Gwennap parish, wherein the mines it served lay, yielded more than a third of the world's copper output at that time and its population of 10,796 was greater than that of Truro, Redruth or Camborne. In the middle years of the eighteenth century the annual production steadily increased and by 1790 had reached 29,000 tons when some ninety mines were selling copper ore at the Redruth ticketings.

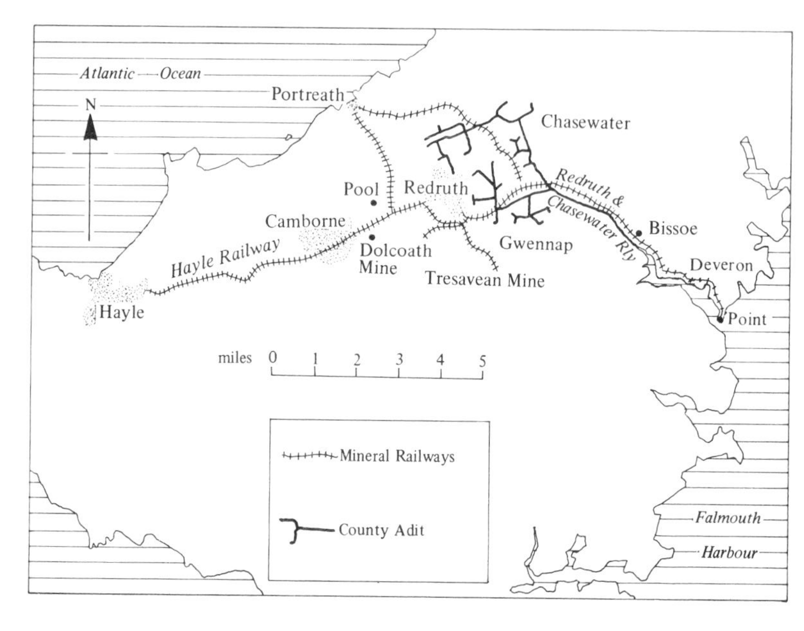

Figure 9 Cornish mineral tramways and county adit

In 1834 the Hayle Railway was formed, the third and largest mine railroad in west Cornwall. It was not completed until 1839. It ran from Hayle harbour past Camborne to Redruth with a branch to isolated Tresavean. From the beginning this line employed locomotives for haulage and at least one of these was assembled at the Copperhouse works in Hayle; successors to the Cornish Copper Co., Harvey's great rivals.

Meantime steam winding engines were being developed for raising ore from the deeper mines. At Dolcoath in 1784 the whim drew 3 cwt of ore from the 180 fathom level but the haulage rope weighed over a ton. Soon after the introduction of Loam's man-engine (see Chapter 4, Tin) wire ropes from Germany came into limited use. South Francis was the first mine to use them.

Blasting became much safer with the invention of the safety fuse by Bickford of Camborne in 1835. In most mines the miners still climbed ladders to and from work. The man-engine was never widely adopted because many of the old shafts were too crooked and narrow and to replace them by a new larger vertical shaft was too great a capital expense for most mines. However, before the end of the century 'gigs' or cages drawn by wire ropes became general for raising men in all the deep mines. Less costly than a man-engine, it was possible to install them in narrow, crooked and inclined shafts.

With these improvements in transport and drainage, copper mining boomed for nearly half a century from 1825. From 1810 onwards adventurers looked for other areas as geologically favourable as Carn Brea and Carn Marth and found one in the strip of hills around St Austell Bay. Though much less important than the great copper-bearing area from Chasewater to Camborne, it was enough to falsify the belief that there was nothing worth mining east of Truro Bridge.

Joseph Thomas Austen (or Trefrey, after marriage), was the sole owner of Fowey Consols which, in 1837, was the second largest copper mine in Cornwall. Par Harbour was built by him and he also built a tramway from the mine and a short canal to the harbour. By means of long leats he supplied sufficient water to operate a number of waterwheels and in this way reduced his coal consumption to less than one-sixth of that required by Consolidated Mines. He was described in 1848 as the biggest employer of labour in the West of England.

In 1836 when Dolcoath became a tin mine, copper mining was first developed in east Cornwall. At Caradon Hill, Peter Clymo and his partners struck lucky in 1836, finding a rich lode of copper ore. The success of South Caradon naturally attracted others from Camborne, as this activity in east Cornwall coincided with the early signs of impoverishment of the Carn Brea copper mines. A standard gauge railway served these east Cornish mines, running to a point below Liskeard and later extended to Looe, which was revived as a port entirely because of the copper traffic.

In the Tamar valley a few small copper mines had been opened up earlier. Old Gunnislake sold ore at the Redruth ticketings as early as 1777. Further east in west Devon were other copper mines, the oldest and largest being Wheal Friendship at Marytavy. In 1796 this mine was managed by John Taylor, then only nineteen years of age. It was 170 fathoms deep and by 1822 it was employing 200 people. The mine was normally worked by a series of thirteen massive waterwheels fed by two leats, one two miles long and the other, from the River Tavy, five miles long. There was an 80-inch steam engine in reserve in case of summer drought or winter frost.

The famous Tavistock canal owed its origin to Wheal Friendship. It was projected in 1803 and took thirteen years to complete, even with John Taylor's driving power. It included a tunnel over a mile long and ended at a double tramway incline down to the wharf at Morwellham, 250 feet below. All the mines in the Tamar valley had the great advantage of being beside a deep valley so that short adits drained them down to a good depth. Also, with such an ample water supply they needed few coal-fed steam engines for pumping.

In the early nineteenth century the sixth Duke of Bedford refused to let the Williams family reopen the shaft at Wheal Maria in Blanchland, near Tavistock, though local miners suspected the existence of promising deposits of copper ore. The Duke was more concerned with his pheasant coverts than any dues. His successor, the seventh Duke, Francis Sackville, was more open to bargaining and in March 1844, Josiah Hugo Hutchins was empowered to work in Blanchland for twenty-one years on payment of one-twelfth dues on all ores. Hutchins and five London stockbrokers formed a joint stock company with 1,024 £1 shares. This new venture was called North Bedford Mines. It began operating on 10 August 1844 with Captain James Phillips as manager. To trace the lode, pits were dug along the probable strike and the shaft was enlarged to 10 ft by 6 ft. The shaft was deepened another 2½ fathoms and on the afternoon shift of November 4 there was a sudden rush of water and the back of the longest and richest copper sulphide lode in the West of England was exposed. By the end of the shift at 10 p.m., £60 worth of 17 per cent copper ore had been brought to the surface. The shaft was called Gard's shaft after Richard Gard, M.P. for Exeter, who held 288 shares. The lode was 40 feet wide in places and the ore brought 'to grass' was dumped in heaps amongst the trees; 192 tons of it sold for £10 18*s* 6*d* a ton.

In 1846 the company changed its name to the Devonshire Great Consolidated Copper Mining Company and issued a bonus dividend of £72,704 without any call on the shareholders as the expenses to March 1845 had only been £2,605. The sett was three miles long by two miles wide and after driving from Wheal Maria for 16 fathoms the lode was lost and panic ensued until it was found several days later shifted 450 feet south by the Great Crosscourse. As the lode continued eastward there were created Wheal Anna Maria (after the seventh Duchess of Bedford), Wheal Josiah (after Hutchins) and, at the east end of the sett, Wheal Emma. This last was named after the mother of William Morris, craftsman, poet and designer, whose father was involved with the company from 1850 to 1880 as a result of having shares assigned to him in payment of a debt. Altogether, there were forty-six miles of levels and it was possible to walk underground for two and a half miles from Wheal Maria to Wheal Emma, there being three miles of underground tramways. The surface works were spread over 140 acres. There were eleven large waterwheels for pumping and winding, leats to supply water eight and a half miles long, twelve main shafts with nearly 2,100 fathoms of pitwork, operated by massive flat rods from waterwheels, and several steam engines. In 1846 the company got permission from the Duchy of Cornwall to draw water from the Tamar for sixteen years for £250 a year. So from three leats the mine got all the water it required for about £5 a week. The Great Leat was two miles long by 18 feet wide and was fed from Latchley Weir. The water from these leats operated machinery at the mine 400 feet above the local level of the river there. Coal from South Wales cost up to 5*s* a ton offloaded at Morwellham Quay.

In the mid-fifties the company was granted a right of way for its railway when transport by road to the river cost 5*s* a ton of ore. The first train ran in November 1859. At the Morwellham incline 3½-ton trucks were attached to a wire rope round a drum operated by a 22-inch sturdy steam engine, the two descending trucks being partly balanced by the two ascending ones. The cost was 1*s* a ton and saved the company £4,000 a year.

Five years after the discovery at Wheal Maria nearly 90,000 tons of ore had been sold, and after all expenses the company gained £300,000, of which the Duke got £44,000 in royalties, and the shareholders £207,000 in dividends. Underground there were solid masses of copper and iron pyrites within a wall of arsenical pyrites 5 to 6 feet thick. The record year was 1857 with 28,826 tons of ore fetching £158,432. The Duke refused offers of large sums for the lease of the setts to the east of Wheal Emma, because he had promised this extension to the company, and in 1857 it was included in the new contract to allow the company to explore along the possible line of the lode for another five furlongs. But he demanded £20,000 for the new lease which shocked the metal mining fraternity, for the Duke had already gained a £100,000 fortune at no risk. He did, however, give Tavistock a town hall, an imposing square and a new market place. The mines consumed 200 tons of coal, 33 cwt of gunpowder, and an enormous quantity of timber each month and paid £1,200 a year in rates and taxes to Tavistock. They had 1,230 employees including 168 girls and 217 boys, some as young as eight years old, hand-picking, breaking ore and tending buddles. School was provided for some of these children on the mine. As late as 1879 the miners' wages were only about £3 12*s* for a five-week month. Dressing floor workers got £3 5*s* for the same period. Unlike the hard-swearing women at other mines, the bal-maidens sang hymns at work, breaking and sorting the ore for a shilling to 15 pence a day.

In 1866 the mine labour was diluted with farm labourers. In desperation the miners formed a Miners' Mutual Benefit Association. On the second of March 1866 the situation was tense and in panic the magistrates sent for troops. The resident director, W. A. Thomas, got the worst of a shouting match, but the Association soon collapsed. In 1878 the miners got greater local support but their wages were cut to 13*s* 7*d* a week, although there was now no five week month. In March 1879 wages were again cut by 10 per cent because the company was losing £5,000 a year. The miners went back on the company's terms in April. Twelve hundred people dined together on the mine to celebrate the twenty-first birthday of the company in 1865, but by 1868 the main lode was nearly exhausted. Wheal Josiah was barren in depth and trials for extensions to the east had proved abortive. At the same time the metal price was low. The 1872 output was little more than half that of 1864. When copper prospects were gloomy the company spent £30,000 looking for tin below the copper belt because it had been found at Dolcoath and other Cornish mines. Richard's shaft was planned to be deepened to 300 fathoms but at 260 fathoms the search was stopped and little more than 2 tons of cassiterite was won.

Peter Watson, a self-made practical metal miner, then took charge and after he had man-engines installed at Wheal Josiah and Wheal Maria, labour relations improved. In the winter of 1887 the Tamar froze over, the price of copper was at its lowest, the Duke more difficult to deal with, and water had risen in the lower levels. A dividend of 2*s* 6*d* in 1899 was the last. In 1900, the Duke was unwilling to renew the lease so by 1901 the mine was moribund; on 30 November all work stopped and the company went into voluntary liquidation. In 1907 all the plant was sold, the Duke had the shafts filled and all buildings levelled to the ground so that there is scanty evidence today of the biggest copper mine in Europe. Six thousand inhabitants of Tavistock depended on the mine for their livelihoods. In fifty-five years it had sold 736,229 tons of copper ore worth £3,473,046. In its last years it had lived almost entirely on the production of white arsenic but that is another story. The Duke gave pensions to workers with forty years service. During the 1920s the brothers Cloke produced precipitate copper from the waters issuing from the mines.

After 1850 it was west Devon and east Cornwall that were responsible for the increased production of the following six years. In 1855 there were twenty-four active mines in the Liskeard area. In the west the ore after 1850 began to be lower in grade; 50,000 tons of ore only produced a thousand tons of copper metal. Nevertheless, some mines south of Carn Brea, that had been regarded as too far south of the mineralized zone, struck rich ore and prospered greatly for a time; 1854 was the peak year for copper production when Devon and Cornwall mined 185,000 tons of over 7.5 per cent copper ore containing 14,100 tons of metal.[8] After the mid-1850s the decline was slow for the first decade and then rapid. The western mines were deep and the grade of ore poor; pumping, hauling and hoisting men over 200 fathoms cost too much for that grade, even though the metal price was high. Some of the best known Camborne and Redruth mines had already become tin mines with an almost unbroken productive life. But tin mining could never absorb the large number of women, boys and girls employed on the surface at copper mines. In any case, tin mining was not flourishing in the 1860s. So the great exodus began; some 7,300 miners emigrated after the 1866 slump and once famous mines not only in Gwennap but all over Cornwall were suspended or abandoned. In 1864 there had been 173 productive copper mines, by 1878, there were only eighty. The great collapse reduced output from 160,000 to 80,000 by 1870, and the metal price fell sharply from £115 in 1860 to £38 in 1890.

By 1875 only three major copper mines survived: Devon Great Consols, South Caradon and Marke Valley, with several small mines producing a thousand tons or more each year. Caradon closed in the 1870s and Marke Valley in 1877. In the west, tramroads were closed and the rails lifted and sold. By 1880, West of England production of copper was only 1/350th part of the world's output. Over 11 million tons of copper ore in all had been produced of which Devon contributed 1½ million. Well over 40 million tons of veinstone were broken and picked on the surface. There were 1,500 miles of underground workings from which some 920,000 tons of metal were produced.

Copper mining in the West of England, or in fact any part of Britain, has ended for all time. Ivy grows on roofless enginehouses; jackdaws nest in stacks that drew the fire from countless tons of good Welsh coal. Heather slowly creeps over the floor and foundations of dressing plants. But the real monument to copper mining will never be seen again: the miles of shafts, the hundreds of miles of levels and the great stopes now drowned, silent and abandoned for ever.

In 1720 the Welsh smelters were few in number and could pay as little as they liked for ore, down from £6 to as little as £2 10*s* a ton. By 1741 all coastwise duties on coal brought into Cornwall for smelting had been abolished and this favoured the setting up of smelting works near the mines. Nevertheless copper smelting needs much more fuel, skill and capital than tin smelting and ventures at Phillack, St Agnes and St Ives did not prosper for any length of time. A venture by Sir Talbot Clarke, who had family connections with Bristol smelters, failed through mismanagement and, situated near St Austell, it was too far from the then productive mines.

South Wales and, to a lesser degree, Bristol, continued to dominate the growing industry and 1725 saw the beginning of the peculiar Cornish system of ticketing to sell the output of the mines. In South Wales there were only six smelters in 1730 and only twelve in 1775. They were closely linked, and conspired to keep the price of ore low by comparing their tickets before bidding at the sales.

In 1755 Sampson Swaine built furnaces at Carn Entral to produce regulus* for shipment to South Wales for refining. He joined John Vivian to form the Cornish Copper Company in a bid to create a viable Cornish smelter. Land was bought on the eastern inlet of Hayle estuary convenient for landing coal and here furnaces, wharves and workmen's houses were built and ships and lighters bought. Vivian was the capitalist and Swaine the metallurgist. The latter had family connections in Welsh Smelting — extremely useful in those days when technical know-how was a jealously guarded secret. The Swansea smelters did all they could to discredit and oppose the Hayle concern, but the firm persevered and was accepted as an established smelter. However, it was too small to become a real rival to the English Copper Co., Freemans, or the Mines Royal. Furnaces erected at Camborne were admirably situated for ore, but the cost of transporting coal by packhorses proved prohibitive. Even at tidewater it was more economic to ship the ore to the coalfield than ship fuel to the ore field. The cost of production at Swansea and Neath was considerably less than at Hayle.

* An intermediate product obtained in smelting ores, consisting chiefly of metallic sulphides.

The coal imports from South Wales for pumping engines were so great that vessels returning to Wales were never in ballast but carried cargoes of ore.

About 4,000 to 6,000 tons of ore a year were smelted at Copperhouse in Hayle, a small throughput compared with Swansea. In conjunction with smelting the Hayle concern started to trade in coal, timber and iron, which was more profitable, lasted much longer than smelting, and led to a great improvement in the navigation of the Hayle river estuary.

As mentioned earlier, the reply of the Swansea smelters to competition from Anglesey was to force down the price of Cornish ore. To defeat this John Vivian and Matthew Boulton formed the Cornish Metal Company with a capital of £500,000. Boulton was convinced that the Cornish copper miners were unfairly treated by the Welsh smelters. The company was to be concerned with marketing only, was to replace 'ticketing' (which Boulton disliked) and make the smelters' agents subservient to mining interests. All went well for a time, but stocks of both ore and metal accumulated owing to opposition from the Swansea smelters and the domination of the market by Anglesey Williams. The Swansea smelters actually imported foreign ores as a counterthreat. Again, unfortunately the Cornish mining adventurers were too individualistic to cooperate; also there was only a market for three-quarters of the output. After long and involved negotiations, North Down and Dolcoath mines were closed down, their shareholders being compensated by an agreed annual sum from the profits of the working mines, but with this reduced production the Gwennap mines were working at a heavy loss. Riots occurred directed against this Cornish Metal Company and also against Boulton and Watt. Because many of the mine adventurers were merchants supplying the mines with all their necessities, these trade profits made it worth their while to continue a nominally losing venture. So, although the riots were misconceived, the adventurers fostered them rather than otherwise. The sufferers, of course, were the cast-off miners who were almost literally starving, as copper sold for so little that mine after mine closed down.

The Cornish Metal Company was doomed, though it had somewhat shaken the Welsh smelters. It had also had a qualified success in selling the copper metal, but its management was disunited. William Williams of Anglesey so dominated the market that he was responsible for selling the company's stocks during the two final years of its existence. It was wound up in 1791, the shareholders being lucky indeed to get their capital back. Copperhouse at Hayle gradually fell into a decline after 1800. With its departure the Cornish copper mine owners abandoned all hope of producing the red metal in their own county, where, of course, they had produced white from black tin for many centuries. They ceased to fight the South Wales smelters' ring which by now controlled three-quarters of the world traffic in copper, and acted in accord with the old adage 'if you can't fight 'em, join 'em'. So in 1830 Pascoe Grenfell went into partnership with Owen Williams, the son of Anglesey Williams, to set up the Copper Bank Smelting Works. He was soon followed by John Vivian who took his Hayle interest to works opposite at Llanelly. He and his two sons managed the works and were backed by several Cornish adventurers. In 1809 the Vivians moved to the larger Hafod Copper Works near Swansea. R.A. Daniell, a wealthy Truro merchant acquired a partnership in the Llanelly Copper Works when it started up. There was a lull until 1831 in the Cornish invasion of the South Wales Smelting industry but then the Scorrier family of Williams and a certain Joseph Foster bought the Birmingham Mining and Copper Company of Swansea. As Williams Foster & Co., they became leaders in smelting with Vivian & Sons not far behind. This Cornish influence in smelting did not mean the betterment of the miners. The smelters continued to combine and their agents went on comparing tickets prior to each sale just as before.

The old Welsh process still remains essentially the basis for smelting sulphide copper ore.[9] The copper reverberatory furnace was developed in South Wales, while Swansea was the largest copper metal producer in the world. The reasons are clear. It was near the biggest producer of copper ore, Cornwall, it had a good harbour and nearby were high grade metallurgical coal deposits. There was also keen competition between the smelters and great secrecy over details of smelter practice. The process was the concentration of copper in a matte from which the metal is separated and subsequently refined. First, there was the preliminary roasting of the ore for the part-elimination of the sulphur content. Then followed the smelting in reverberatory furnaces to concentrate the copper in a matte and to eliminate the gangue as slag. Finally, the bessemerizing of the matte to produce blister copper and the refining thereof.

The pre-roasting is not only to burn off part of the sulphur but also to oxidize the iron so that it combines with the silica in the reverberatory furnace to produce a fluid slag. Copper has a greater affinity for sulphur than other metals. In the smelting process the copper forms cuprous sulphide (Cu_2S). High grade matte with all the iron oxidized is most desirable, but many cost factors were involved which the individual smelter had to weigh. The higher the grade of the matte, the greater the copper content of the slag and also the greater the copper losses in the form of dust from the furnaces.

The atmosphere in the reverberatory furnace being neutral the ferric oxides are reduced to the ferrous necessary for combining with silica, which is performed by the sulphide of iron — as in the following formula: $3Fe_2O_3 + FeS = 7FeO + SO_2$, the FeO forming slag with silica.

The speed of reaction on the copper depends on the sulphur left after roasting. Thus if the charge is low in sulphur it is more refractory in smelting than one high in sulphur. The matte formed in the reverberatory furnace is not a chemical composition but a mixture of cuprous and ferrous sulphides. The copper is isolated in the matte by progressive oxidation in a refractory-lined vessel called a converter. When air is blown into molten matte the iron sulphide is oxidized before the copper sulphide. Further blowing converts the copper sulphide to copper. Heat is supplied by the oxidization of the iron and copper sulphide, the reaction being highly exothermic. As iron sulphide supplies most of the heat, the amount of it is important in controlling this reaction.

Refining the crude copper by fire methods sometimes removes the bulk of the impurities, but today it is generally only a preliminary to electrolysis, the present universal method of refining. Fire refining does not effectively remove precious metal nor deleterious metals like bismuth and selenium. Electrolysis removes such contamination and also makes possible the recovery of gold and silver.

In earlier times when much high-grade coarse ore was available the blast furnace was prominent in reducing copper ores. Because the blast furnace is flexible and has a low cost of maintenance it was then a serious rival to the reverberatory furnace. The sulphide ore was first roasted and the furnace charged with roasted ore, coke and fluxes and the copper reduced to metal, but the copper loss in the slag was high. The blast furnace has lost its former importance and the reverberatory is better able to cope with the finely divided flotation concentrates which form the bulk of the supply of copper today. In the old South Wales practice about 1800, a 12-foot long reverberatory furnace was normal. At Anaconda in 1910 a 140-foot furnace was in use though the limit now is about 110 feet, with better firing and the use of pulverized coal. A century or more ago the proportions in a charge of 10 tons of ore were 1 ton of fuel to 1 ton of ore. Now with a charge of 2,000 tons they are 1 ton of fuel to 10 tons of ore using improved refractory linings and continuous tapping.

In the converter operation, the total time in one typical case was 11 hours 13 minutes with 86.2 per cent blowing time and an output of 62 tons of blister copper for an input of 150 tons of matte. Recent improvements in copper smelting technique are not touched on here as there has been no copper smelter for many years in the United Kingdom. Today of course, electrolytic refining is a *sine qua non* for primary copper. The intrusion of Cornishmen into Welsh copper smelting coincided with the beginning of Cornish interest in overseas copper mines. Ticketing of foreign ore in Swansea began in 1804 on the Redruth pattern. The ore came from Mexico, Cuba, Peru and Chile but not until 1820 did the earliest large shipments reach Swansea. It was the first tiny but ominous crack in Cornish copper monopoly. The American smelting of their native ores destroyed the Swansea monopoly where the industry steadily declined to become nearly extinct by World War I.

CHAPTER SIX

Lead

It is generally accepted that our lead deposits were formed by mineralizing solutions ascending through fissures.[1] The commercially important deposits in Britain fall into two groups: those in carboniferous rocks and those in precarboniferous rocks ranging from Silurian and Ordovician to rocks of doubtful Cambrian age. No worthwhile lead ore deposits are found in rocks younger than the carboniferous age. Commercially important lead deposits occurred in the carboniferous limestone series and in some cases in the beds immediately above, normally in steep dipping lodes or 'flats' of greater lateral than vertical extent, usually under a substantial sealing bed of shale.[2]

Geographically, the precarboniferous deposits are confined to the western half of Britain and the carboniferous deposits to the Pennine chain and its flanks, and to Flintshire, Denbighshire and the Mendips of Somerset. In Derbyshire until the late 1930s it was thought that lead ore was cut off by the Toadstone*, but then Leslie Williams's staff found highly productive orebodies under this amygdaloidal basalt deposited contemporaneously with the limestone.[3]

In Britain a greater tonnage of lead has been produced than of any other base metal. A Government report of July 1949 gave the total of output over the years 1850 to 1949 as 4.5 million tons of concentrates averaging 75.8 per cent lead yielding 3.3 million tons of metal.[4] Before that the Beaumont and London Lead Companies from about 1700 to before 1850 produced $1\frac{1}{4}$ million tons of dressed lead ore of about 70 per cent lead yielding between 0.8 and 0.9 million tons of metal.[5] As many other lead mining fields were active during and before that same period, a conservative estimate of the total from 1700 onwards to date is 5 million tons of metal. Though little lead was produced after 1950, previously,

* German *Todtstein:* dead rock.

Figure 10 Lead deposits

in Roman and medieval times, there were periods of vigorous exploitation. So, throughout its history, this island has produced at least 6 million tons of lead metal and a strip of country in North Wales only twenty-five miles by three produced £20 million worth of concentrates since the beginning of the eighteenth century.[6] The period of greatest production was from the Napoleonic Wars to the 1880s, reaching a peak between 1854 and 1871 when the annual output only once fell below 90,000 tons of dressed lead ore. In 1856, 99,514 tons of lead ore was won from 347 mines[7] in Great Britain.

In the brief revival from the 1930s to 1944 three mines, Mill Close, Halkyn and Greenside, in widely different parts of the country, produced about 40,000 tons a year of contained metal.[8] Halkyn went out of production in 1941, Mill Close in 1944, and Greenside in 1948 produced 2,500 tons of the country's small total of 3,000 tons.

The pronounced decline since 1880 has been ascribed to several causes: crippling taxation, wide fluctuation in international metal prices and competition from overseas lead mining concerns.

The total ouput of lead concentrates from 1845 to 1938 in Great Britain divided into geological horizons was:

	Tons lead concentrate
Upper Palaeozoic rocks (mainly carboniferous limestone)	3,082,455
Amorican (Devon and Corrnwall)	316,967
Lower Palaeozoic rocks of Cambrian, Ordovician and Silurian age	1,563,941
Trias	520
Total	**4,963,883**

During the last two decades annual output has varied from 2,000 to almost 10,000 tons of lead concentrates. In the last few years production has been at a steadier five thousand tons, nearly all a byproduct of fluorspar mining at Laporte's Cavendish Mill at Stoney Middleton, Derbyshire and at the Weardale Lead Company's works.[9]

Even in the latter part of the Bronze Age tools had some lead in the alloy and a Bronze Age socketed celt of lead was found in a cave near Stanhope in Weardale. Plenty of leaden objects from the later Iron Age have been found, such as cast spindle whorls for use in weaving. Nevertheless the pre-Roman working of lead ores in Britain was slight.

In a glance back to the opening chapter regarding the status of the native miners, mainly lead miners, under Roman rule, Tacitus wrote that the British submitted, 'to the levying of troops, to the tribute and the other charges of the Empire with cheerful readiness provided there were no abuses. These they bitterly resent, for they are broken into obedience, not to slavery.'[10]

Historians agree that some of the natives employed in the mines were not slaves but voluntary labour, probably most of them working on shallow outcrops. There were also penal settlements where prisoners and political offenders were sent to work in the mines and perhaps kept underground for considerable periods, as the evidence of Roman finds suggest. The Elder Pliny's remarks on the case of lead mining in Britain probably referred to the Derbyshire Rakes which the Romans exploited fully. In the turmoil of wars between Britons and Anglo-Saxons, English and Danes, lead mining was neglected although the Danes worked lead mines in Derbyshire.[11] Later, the Doomsday Book records seven lead mines in Derbyshire. Some of the old terms used in mining are of Anglo-Danish origin such as 'wrowt' for excavating the ore and 'greaving' lead from the old Norse verb 'grafa', to dig.

During the twelfth century the output of lead was relatively large from Northumberland, Durham, Shropshire, Derby, and the Mendips, the demand coming from the builders of churches, castles and monasteries in order to procure a watertight roof. A big boost to Britain's metal mining in general was given by William Cecil, Elizabeth I's minister, whose foresight and perspicacity saw the need to bring in German miners and smelters, who had then long been technically the most advanced in the world. The next great step forward in lead mining was the formation of the joint stock London Lead (or Quaker) Company during the reign of Charles II. It became the largest and most technically advanced lead and silver producer in Europe. Some references have been made to its widespread activities in the chapter on silver. Before settling in the North of England they were active at Leadhills, North Wales, Central Wales and Derbyshire. They even sent men to the Orkneys to make trials.[12]

Lead ore is widely distributed in Scotland and veins occur in most of the counties. At some time or another many deposits have been worked, but mostly on a small scale and to only a shallow depth. During the nineteenth century lead mining was actively pursued in Argyllshire, at Tyndrum in Perthshire, near Newton Stewart in Kirkcudbright and most importantly at Leadhills and Wanlockhead on the borders of Dumfriesshire and Lanarkshire, one of the biggest British lead-mining fields, and God's Treasure House in Scotland.[13] Bronze and stone tools found in old surface workings there suggest exploitation in pre-Roman times. Whether the Romans operated lead mines at Wanlockhead and Leadhills is open to doubt, though pigs of Roman lead metal have been found in the area. Lead mines on the Isle of Islay were probably worked by Norsemen when they occupied a large part of Scotland. About A.D. 1300 lead mines were working in West Argyll when the Constable of Tarbert's accounts mention a payment of 12 pence for charcoal and wages for smelting lead ore. At Glenorchy in Argyll in 1429 a lead mine was declared a royal mine when the Scottish parliament granted to the King all mines yielding more than three ha'pence of silver to the pound of lead metal. Often the dressed ore was exported and between 1585 and 1590, 15,717 stones of lead ore were exported from Leith to be desilvered.

At Leadhills, the highest village in Scotland at 1,350 feet above sea level, the first record of mining in the district is of the monks of New-battle working a mine at Glengonnar in 1239, but there was lead mining there long before that date. When the gold rush was on, lead mining was abandoned and lead metal actually imported to refine the gold. In 1562 lead mining commenced again and that same year Thomas Foullis got possession of these mines and employed Bevis Bulmer (mentioned in the chapter on gold), who abandoned lead mining for alluvial gold. Thomas Foullis's daughter Anne, an heiress, married James Hope of Hopetoun, the ancestor of the Marquis of Linlithgow, the proprietor, who himself worked the mines for a number of years, after which the lease was granted to various companies. In 1735 the Scots Mining Company secured as manager Sterling the mathematician, who divided the workers into four classes, miners, labourers, washers and smelters; he also started the peculiar land tenure system of the district. After twenty years of litigation the Scots Mining Company relinquished their lease and the Leadhills company took over and only shut down in 1929 after two centuries of continuous operation. Leadhills was the birthplace of both Allan Ramsey and William Symington whose fathers were mine managers there in the eighteenth century.

The district is in an area of intensely folded lower Silurian rocks along a north-east to south-west axis which show an extraordinary repetition in outcrops. Signs of repeated movement were seen in the slickensided galena and blende.

Wanlockhead mine worked from 1264 to 1934 with only a few breaks. Its lead veins are said to have been rediscovered by a German, Cornelius Hardskins, during James VI's minority. The mine was worked with fair success up to 1691 by James Stampfield. Matthew Wilson succeeded him and held the lease till 1710 and was himself succeeded by the 'Quaker' company, which in 1721 joined a miners' cooperative called the Friendly Mining Society, to work all the principal veins extensively until 1727 when the partnership was dissolved. The leases were surrendered in 1734. After several changes of ownership, the Duke of Buccleuch took over and worked the mines with moderate success until 1806 when they passed to the Wanlockhead Lead Mining Company.[14]

Wanlockhead's main shaft and works were at an elevation of 1,195 feet, just below Wanlockhead village. By 1920 the shaft was 140 feet below sea level, the mine being worked to a depth of 1,335 feet. It continued to be worked until 1934 when it became flooded and was abandoned. Following a geological examination in 1952, however, the mine was dewatered in 1954. In 1955 a dressing plant was built and milling started in May 1957 at the rate of 200 tons a day. But by May 1958 the mine was on a care and maintenance footing, and in May 1958 it was again abandoned.

Although the Leadhills and Wanlockhead vein system is one, the mines are in two different counties, Lanarkshire and Dumfriesshire, and there was never any underground connection nor any managerial link between the two. They had separate adits, mills, smelters and offices, and many interesting relics remain to be seen in the neighbourhood.[15] In the graveyard of Leadhills, for example, there is a tombstone recording the extraordinary case of John Taylor (1637-1770) who died at the age of 133 having worked for over a century in the lead mines.

In Galloway the most productive lead veins were the West and East Blackcraig mines situated two miles south-east of Newton Stewart. They were discovered in 1763 by a soldier who was helping to build a military road. Soon afterwards the two mines were opened up and the ore sent to Chester for smelting. An English company ceased working there in 1839 and dismantled both its dressing and smelting plants. In 1850 these mines were reopened and worked on a considerable scale until abandoned in 1880.

Another productive venture was the Cairnsmore mine, 2¼ miles from Creetown Station. It was worked on a small scale until 1845 when it was leased to the Kirkcudbrightshire Mining Company which produced some 3,280 tons of lead before 1855 when the company sold the property to a Mr Thomas Field for £550. Ore raised between 1847 and 1855 sold for £36,000. The mine was abandoned in 1859.

The third productive mine in the area was Woodhead, about three miles west of Carsphairn. Discovered in 1838 it started as an opencast from which several hundred tons of lead ore were won. When it was worked underground a large dressing plant and a smelter were erected. The output rose to over 900 tons in 1842 but then steadily dwindled. Only lead ore was mined.

On the Isle of Islay many trials were made for lead ore. The veins occur in the stretch of limestone from Portcraig in the north-east to Bridgend in the south-west, the most mineralized area being in slightly metamorphosed rocks near Ballygrant where there are traces of early workings. Lead was mined there by Norsemen, but the first authentic record is in 1549 when the Dean of the Isles, Donald Hurd, wrote 'In Illa is Meikle Lead Ore in Mochylls'. In 1616 the grant to mine copper and lead in Islay was made to Archibald Primrose, but from 1720 it was in the hands of the Glasgow Company, after which it changed hands a number of times. When the rebellion of 1745 broke out the lessee, Captain William Thynne, returned to England. Pennant, in his tour, visited the island and condemned the work done by the Glasgow Company but declared he never saw so many good veins so close together. About 1770 the Leadhills manager, Alexander Sheriff, visited Islay and made a comparison of costs showing the island's costs to be almost half those of Leadhills.

In a remote part of Argyllshire near the head of Loch Sunart on the north side lies the village of Strontian and the river and valley of that name. Lead mines were discovered in 1722 by the owner of the estate, Sir Alexander Murray, who worked them himself for a few years and in 1729 leased them to the Duke of Norfolk and Company for one-sixth royalty, but soon after the mines came into the hands of the York Building Company who worked them continuously till 1875. Houses were built for the miners together with malt kilns and brewhouse, but the venture shut down suddenly for reasons unknown. Interesting remains of furnaces and houses can still be seen. It was here in the 1790s that the new element strontium was first identified.

On the borders of Argyll and Perthshire is the lead district of Tyndrum and Coninish. The Tyndrum veins were discovered accidentally in 1741 by Sir Robert Clinton from Nottingham who was working a lead mine in Lorne. He got a lease from the Breadalbane estates in 1730 and raised 1,697 tons of lead ore between 1741 and 1745. He appears to have been a Jacobite and his works were occupied by Government troops who stole lead and damaged the plant. Later the property passed through several hands, until 1768 when it was the property of the Scots Mining Company which put up a new smelter and raised 3,685 tons of ore. Then in 1858 it was taken over by the proprietor, the Marquis of Breadalbane, until he died in 1862.

In the Orkneys there were two mines on the shores of Hoy Sound. The one on Hoy was known since before 1529 from a description in latin by Jo Ben, thought to be John Bellendon, Canon of Ross, who said lead ore was mined there. The other Stromness mine on the mainland was small and shortlived although in 1755 English miners sent away a considerable tonnage of ore.

Like all Quaker industrial concerns the London Lead Company was an exemplary and enlightened employer and had the welfare of all its workpeople men, women and children, at heart. Through mixed motives of philanthropy and self-interest they aimed to maintain an efficient, well-fed, well-housed, healthy, sober and contented labour force. They always fostered promotion from the ranks and had no desire to reduce the miners to a state of cringing dependence.

Many times during the first century of their work in the north Pennines, when bad winters with deep snow caused acute distress, the Company sent food supplies to Alston to be sold at London prices. In 1793 prices rose on the outbreak of war with France and in 1795, when there was dear bread in the north, the company's agents were authorized to buy grain in Newcastle market and sell it to the miners at cost price. In 1816 Robert Stagg raised the monthly advance from 30*s* to 40*s* when wheat and rye in quantities were purchased and sold to the miners at less than cost. In the depression after 1815 many of the small mines closed and their workmen had to move or go on relief, but, as the secretary of Greenwich Hospital reported in 1823: 'The London Lead Company continued the working of their mines at a heavy loss. . . . They have proved the great support of the labouring people in the period of difficulty, not proportioning the rate of wage . . . by the standard of the lead market but by the cost of provisions.' The company provided reading rooms and libraries at all its main centres and even at lodging houses. These lodgings, for men living at some distance from the mines, were furnished with bedsteads, often in two tiers with chaff mattresses, the miners finding their own bedding. A room 18 by 15 feet would hold twenty-eight men, and as these dormitories were unheated in winter the men blocked the ventilators. In 1842 a sub-commissioner, Mitchell, reported 'The smell was utterly intolerable. I should think it no hardship to . . . remain 24 hours in a mine but I would be terrified . . . to be shut up for a quarter of an hour in the bedroom of a lodging shop.'

However, in the 1820s the company built many four-roomed cottages to let at £4 a year, and after the outbreak of cholera in 1849 the company spent a considerable sum on reservoirs and piped water at Nenthead, Milton, Dufton, Carrigill, Stanhope and Middleton-in-Teesdale.

The social history of this remarkable Quaker company embraced a guaranteed minimum wage paid monthly, ready money shops, no credit trading, subsidized food supplies in times of distress, a cooperative corn association, smallholdings, free medical care, social security fund, libraries, old age pensions and superannuation. At schools where the salary recommended for a primary school teacher by the Brougham Committee in 1818 was £24 the 'Quaker' company paid from £80 to £100. For drunkenness there was first a warning, then a fine, and dismissal at the third offence. The company even purchased large tracts of moorland so that their employees could hunt small game to supplement their diet and to satisfy their poaching instincts. This welfare work started two centuries ago.[16]

The other great mining concern in the North Pennines, the Beaumont company, operating fifty mines, was the only one to follow the London Lead Company's example. Under Thomas Sopwith it built four schools on the mines at Allenheads, Carshields, Brideshill and Newhouse; children paid sixpence a month and found their own books. Before 1860 Sopwith created libraries at Allenheads, Allen Mill, Coalcleugh and in Weardale. In 1842 the large smelting concern on Alston Moor, the Greenwich Hospital, paid £80 a year towards the support of seven schools on the moor.

Sir William Blackett bought Allendale from the Fenwicks of Wallington, in 1694,[17] and leased all the Bishop of Durham's mines in Weardale in 1698 at a royalty of one-ninth. His daughter married Thomas Richard Beaumont whose grandson, Wentworth Blackett Beaumont, transferred the northern property to his eldest son, Lord Allendale. It was a private family company. The manager in 1845, Thomas Sopwith, increased the bargain rates to give an average wage of 15*s* a week. He also raised the monthly allowance to 40*s* to qualify for which a miner had to work 8 hours for five days a week. In 1855 the Beaumont miners were better off than those of the London Lead Company with earnings of 19*s* a week. In both companies the regular monthly advance was really a guaranteed minimum wage and was normally sufficient for fuel, food and rent for a frugal family, especially if supplemented by the produce of a smallholding.

The mining field worked by these two powerful companies comprised parts of Cumberland, Northumberland, County Durham, Yorkshire and Westmorland, an area of 650 square miles, with its southern boundary from Scotch Corner to Brough and on the north the line of the Roman Wall. There was some mining activity as far back as the twelfth century but there is no statistical information before 1666, and from 1700 onwards the industry was dominated by the two large companies.

In 1884 the Beaumont family concern surrendered its leases after nearly two centuries of productive mining and both leases and plant were taken over by the Weardale Lead Company. Some Allendale leases were added much later in 1924. This new company carried out well-planned development, found new high grade deposits and worked on until 1940. It was lucky with the Boltsburn mine where large rich 'flats' were found which enabled the company to continue working profitably.

In 1882 the London Lead Company sold its plant and leases to the Tynedale & Nenthead Zinc Company, which re-sold them to the Vieille Montagne Company of Belgium in 1896. This latter company opened up a rich new development area at Nentsberry Haggs Horse Level producing over 40,000 tons of concentrates of 80 per cent lead. Another of its finds was the Rotherhopefell Mine.

Boat-shaped pigs of lead were still being cast in Weardale in 1908, a practice going back to the days when all lead was transported on packhorses to Newcastle for shipment. Two pigs were a single horse's load and they were shaped to fit the animal, being held in place by ropes.

In Yorkshire, lead mining for all practical purposes is divided into two areas with entirely separate development but with considerable migration between them, especially from Swaledale to Grassington. One area is Swaledale, Arkengarthdale and Wensleydale the other Greenhow Hill and Wharfedale. In both, the evidence of mining starts with Roman pigs of lead.[18] There is no mention of mining in the Doomsday Survey, but in the later part of the twelfth century a group of miners were operating in Richmondshire. Roger de Mowbray granted usage of large areas in upper Nidderdale to the Cistercian monasteries of Fountains and Byland because he wanted to set out for the Crusades with his son Nigel, and the Lord Abbot of Fountains donated 120 marks to him and 10 marks to Nigel in aid of their journey to Jerusalem in 1375.

There are only scanty records until the sixteenth century. The monks of Bolton Priory worked mines in Appletreewick in a small way and Swaledale and Greenhow Hill were operated directly or indirectly by the monasteries until the Dissolution when they reverted to the Crown, which leased them to various notables. In 1533 Henry VIII granted to Sir James Metcalfe the working of the mines of Middleham and Richmond and before 1590 a smelt mill was built at Marske.

Charles I, in recognition of an enormous loan of £230,000 granted the lordships of Middleham and Richmond to Henry Ditchfield and Humphrey Clarke. In 1603 Derbyshire miners were brought to open up mines at Yarnbury, but after a steady loss all the mines were leased in 1620 to independent miners, the smelt mill being kept by the 'lord of the field', the Earl of Cumberland. By the 1630s the mines were showing profits. The Earl took a third of the smelted lead as his dues and his profit in 1638 was £212.19*s*.

The two most important mines in Swaledale, West Riding, were Old Gang (see Plate 23) and Lownathwaite. Philip Swale, a lawyer and Lord Wharton's steward, formed a partnership with Robert Barker of Richmond in 1676. They took a lease of a part of the Old Gang and Lownathwaite mines for twelve years at 22*s* a fother* of ore won. The lease was renewed, and by 1700 the output was averaging 440 tons of lead a year. While these mines were being developed 'Old mens' working were found, proving that there had been workings there in Roman times. The mines prospered and a profit of £5,117 was gained. Swale, a progressive manager, brought in Robert Barker's brother Adam from Derbyshire when Robert died in 1680. Adam introduced bellows to improve ventilation underground. When leased in 1771 Old Gang had a duty of one-fifth plus a rental of £2,163 a year calculated as interest on the capital invested by the proprietor in driving horse-levels. These were adit crossents, sufficiently large for a pony and a rake of trucks to draw out the ore; they were also used for drainage. This was the highest rate known for a lead mine lease.

* A fother varied from 2505½ lb (York) to 2184 lb (London).

In 1811 Old Gang mine was leased to George and Thomas Alderson, two London lead merchants, and they agreed to drive 1,500 yards of horse-levels in four stretches. John Davies, agent for the Pomfret-Denys mines from 1802 to 1822, was made manager and given a free hand, the Aldersons only infrequently visiting the mines themselves. Davies was not only dishonest but incompetent. Under his direction the Bunting level was driven to connect with the Hard Level, but when it reached the point of junction it was 24 feet too high. Within three years of his appointment there was a loss of £20,000 so he was dismissed in 1814.

His place was taken by Frederick Hall, who quickly introduced inclined planes, steam engines and cast iron rails in the horse-levels instead of wooden ones. Davies disliked Hall as much as Hall despised Davies. From about 1790, as a result of well-spent capital, the output rose from an average of 583 tons a year from 1786 to 1789 to over 2,000 tons from 1800 to 1809. In 1801 a rich strike brought the output up to 3,252 tons for that year. For thirty years from 1830 this company's total profit was estimated at £100,000.

The partnerships at Old Gang were paid monthly. When the pickmen were working the rich Watersykes veins their earnings were the highest in the company's records. In November 1872 it was as much as 103*s* 2*d*, whereas in June 1873 it was as low as 6*s* 9*d* because many miners only put in short time while haymaking. The great depression in lead mining leading to the closure of many mines in 1880-82 forced wages down to starvation level. At Old Gang pickmen took up less than 9*s* a week in the winter of 1884-85. The slump also caused migration from the mining towns. The population of Melbecks, where Old Gang Mine was situated, fell by 58 per cent between 1871 and 1891.

A Newcastle firm which manufactured red lead leased the whole Arkengarthdale field in 1800. These mines were poor in the early 1770s but successful trials in virgin ground revived their fortunes later in that decade and new discoveries in the Danby Stang and Faggergill mines kept them profitable in the early 1790s in spite of low metal prices. These mines made no regular subsistence payments but did allow credit, while advances were made both in money and kind. The food was drawn at the company's shops, the advances being repaid from earnings provided they covered them.

The decline in mining at Arkengarthdale was later than at Old Gang but it lost the same 58 per cent of its population between 1881 and 1901.

The Derbyshire Rakes in the northern half of the Derbyshire minefield are outcrops extending in some cases for miles in a country bare of glacial drift and even soil. They were so easily recognized that they must have been worked by the Ancient Britons long before the Romans came. Some writers put this date at 300 B.C., but it must be remembered that the knowledge of making metals came to Britain early in the second millennium B.C. through the Beaker folk.

The earliest method of working these outcrops was trenching along the vein at the least possible width to allow extraction of all the ore (see Plate 20). The working would have been taken down some 20 feet until the sides of the trench became unsafe or water filled the trench or both. Later, small shafts were sunk along the vein down to a depth of 30 feet or more; tunnels were then driven along the vein, as far as safety and ventilation allowed, to take out as much ore as possible before sinking another shaft and repeating the process. So the line of the vein came to be marked out by a line of shafts like a string of beads. If a vein or lode was wide enough it would be excavated on all sides at the bottom of the shaft to form a 'bell pit'. The early work in Derbyshire was in this case little different from the method of extraction used by the neolithic flint miners at Grimes Graves in Norfolk, except that instead of stag's antlers and blade bones of oxen, the Derbyshire miner used iron tools.

As noted earlier, tradition arose in Derbyshire that no ore would be found below the 'toadstones'. This was disproved in recent years in a spectacular manner at Mill Close. But before that many of the ancient outcrop workings were below toadstones as Farey's important record showed when he listed 271 operating mines in 1811. When the Derbyshire mines grew deeper to some 50 fathoms the shafts were not sunk directly down to the bottom but for about 50 to 60 feet only and then other, similar shafts were sunk some 6 to 7 yards to one side. This system of descent by a series of short shafts was a widespread practice as long as ore was raised by muscle-power, one man winding up the filled kibble by a windlass at the top of each short shaft. The miners could climb up on 'stemples'—short timbers set into a corner of the shaft to form a primitive ladder. When such a shaft was stonelined, 'pigeon holes' were left in the stonework to enable the miners to get a foothold and climb the shaft. The diameters of the shafts were so small that danger of falling was slight (see Plate 19). These climbing shafts were in use up to the nineteenth century, sometimes from considerable depths.

The minerals of the Derbyshire veins are galena and blende (sulphide of zinc, sphalerite), their oxidation products, and fluorspar, calcite and barytes. The old miners thought that lead would be found wherever fluorite occurred.

The Derbyshire mines were in two areas, the King's Field or High Peak and the Wirksworth Wapontake or Low Peak, both in the possession of the Crown through the Duchy of Lancaster which leased the mines to secular workers. Few of the earlier grants imposed a royalty payment based on production; most were rented at a fixed rate. During the thirteenth century Crown lawyers regarded the Peak mines as royal property, but the right to work them was governed by the miners' customs. In 1287 the miners appealed to the king to restore their ancient liberties and a commission decided in 1288 in favour of the miners. Nevertheless these immemorial customs were not codified until the Derbyshire Mining Customs and Mineral Courts Act of 1852 was passed by Parliament.

The privileges claimed by the miners were that a miner on finding a mine should be granted two meers* of land, one a reward for finding and the other secured by payment of a dish of the first ore won. For royalty the king had the thirteenth dish of ore. The miner enjoyed right of access from the nearest highway. The king had the first option of buying the ore won, but he had to pay as much as anyone else. The miner could sell his meer without the barmaster's consent. Small Barmote Courts were held every three weeks at the mines. Great Barmote Courts were held twice yearly with a jury of twenty-four men knowledgeable in mining customs. The miners also had the right to cut wood for timbering the mine, for smelting the ore, and for possession of the mine by erecting wooden stowes (headgear); also the right to secure the hand of an offender to the stowe by a knife as a penalty for a third theft. The Barmaster, who was appointed by the Lord of the Field, in this case the Duchy of Lancaster, was the most important official; he measured the meers allotted to a finder, kept a record of the ore produced, collected the lord's 'dish' and summoned the regular Barmote Courts. He was required to visit the mines regularly, noting those being worked and seeing that the mining was carried out in a proper manner and in accordance with the customs.[19]

In the fourteenth and fifteenth centuries a large number of small Derbyshire mines washed and dressed their ore in local streams and lead

* In the High Peak a meer was 32 yards, in Wirksworth 29 yards.

merchants bought small parcels at fairly regular intervals providing the poor miners with some ready cash. The men and boys worked in the mines, and the women and girls dressed the ore. The late fifteenth and the sixteenth century were generally prosperous, mining being financed by merchant-smelters who included some of the landed gentry. John Manners of Haddon Hall dealt in lead and in 1578 asked the Earl of Shrewsbury for permission to continue to use a foot-blast for making lead. Haddon Hall was roofed with soft Derbyshire lead ingots beaten into sheets.

The earliest Derby sough was started in 1633 at Teargal mine near Wensley, where work had been stopped by flooding; by 1635 this sough had dewatered many working places permanently. The important seventeenth century Dovegang mine was soon worked out down to water level, so the owners called in a famous Dutch engineer, Sir Cornelius Vermuyden, to make a drainage adit. This sough was driven between 1629 and 1636 and gave many more years of life to the mine which, in 1652, paid over half of the king's royalty in the Wirksworth field. In the eighteenth century there was a period of success until a decline set in by 1790 when most mines were losing money. By the turn of the century many mines had closed and by 1807 all were shut down during the post-Waterloo depression. Nevertheless, throughout the rest of the nineteenth century Derbyshire was a very active lead mining field with a production comparable to that of Alston Moor until about the 1860s.

The complexity of the small mine and sough companies made systematic mine planning difficult. Only a big company like the Gregory mines could afford steam engines working from the surface for pumping. Generally the mines were served by horse whims of limited capacity. There was never anything in Derbyshire comparable to the Beaumont, London Lead or even the Old Gang companies in the North. The Gregory and Ashover group closed in 1856 and the Alport mines in 1853, leaving some sixty small mines worked by thirty companies.

In 1856, when Great Britain produced nearly 100,000 tons of dressed ore, Derbyshire produced 9,534½ tons or nearly a tenth, only capped by the north Pennines, Yorkshire and Cornwall.[20]

The most important mine in Derbyshire was Mill Close. In 1741 the London Lead Company purchased the mine for 1,000 guineas. In the same year the Yatestoop Sough was begun by the London Lead Company and other interested parties who were to benefit. It was completed in 1764 at a cost of £30,000. In 1742 the company cut the vein at shallow depth, so a 42-inch cylinder steam engine from Darby of Coalbrookdale was installed in 1748 to raise water to the sough level. It lifted water 144 feet from below the water level in the main shaft and by a complicated system of rods and cranks operated a second pump in another shaft which lifted water 150 feet to sough level. The success of the Yatestoop Sough inspired others, but few enjoyed such good returns, and Farey in 1811 reported that they were mostly unprofitable speculations because they took so long to drive that meantime the miners by pumps and short soughs had continued to mine the ore. Nevertheless, except for Mill Close and the Gregory mine at Ashover, there was not much mining below the level of the great soughs. Mill Close continued to be worked by the London Lead Company until 1778 when that 'Quaker' Company severed all its links with Derbyshire. Thereafter the mine lay neglected until 1853 when it was reopened by a Mr Wass who, after two years, found good ore and kept up a steady output. In 1870 Mill Close produced half of Derbyshire's total output. Wass died in 1886 and his trustees carried on mining at Mill Close until 1919. In 1922 the mine was sold to Millclose Mines Ltd, which got good ore from 'flats' and cavern fillings. In 1929 the mine was worked down below the upper limestone and rich ore was found in successive steps northward under the toadstones. Between 1929 and 1939 the workings were deepened to over a thousand feet below the shaft collar. As much ore was gained under the toadstones as the whole previous output above them. In 1939 the base of the oreshoot was reached and that, with the incidence of World War II, led to the closure of the mine in 1943. An added problem was the increasing cost of pumping the growing volume of incoming water (see Plates 25 and 26). During its long life this mine produced half a million tons of lead concentrates and its output during the 1930s was phenomenal.

In proportion to its size, the output of mineral wealth of the Isle of Man has been enormous. Veins are exposed in the cliffs; for example, the great lode of Bradda Head, with its gangue of white vein quartz cutting vertically through the dark slate cliffs, is so conspicuous from land and sea that it must have attracted attention as soon as man knew metal.

The first mention of the Manx Mines was in 1246 when the island was a Norwegian dependency and King Harold II gave the monks of Furness Abbey the right to work the mines. Later, John Comyn, Earl of

Buchan got a licence from Edward I to dig for lead to roof the eight towers of his Cruggleton Castle in Galloway. In 1406 mines of lead and iron were included in the grant of the island to Sir John Stanley by Henry IV. In the mid-seventeenth century Captain E. Christian found that the lead ore at Bradda Head contained much silver. In 1688 a lease of all the mines on the island was granted to two merchants, one from Liverpool the other from London, with leave to erect one or more smelting mills. The first recorded smelter on the island was built in 1711.

Charles, Earl of Derby, was granted a lease of all Mines Royal of gold or silver 'holding such a proportion of these metals as according to the Laws of the Realm of England makes the same a Mine Royal'. This lease expired in 1735 but was revived on a petition of John, Duke of Athol in 1780 upon a declaration made by a former Deemster, P.I. Heywood, who declared that there were no mines of gold or silver, and that the only mines ever worked were mines of lead and copper. Also that there was a proportion of silver in the lead mines now working but too small to answer the expense of assaying and separating'. This was obviously an attempt to sidestep the restrictions of the oppressive Mines Royal Act.

In 1811 Woods found Foxdale mine deserted and drowned, Laxey being worked on two levels only and yielding silver-lead and copper ores. A work published in 1819 noted that all the mines on the island were abandoned, with no prospect of reopening. But from 1845 the company working the Foxdale group employed 350 men and boys and raised about 2,400 tons of silver-lead ore a year. Laxey was worked by a new company which employed 300 men and raised 60 tons of lead ore, 200 tons of mixed blende and galena and 5 tons of copper ore a month. For the next thirty years both Foxdale and Laxey enjoyed great prosperity. Great Laxey mine produced from official records 64,116 tons of lead ore, and 8,633 tons of copper ore from 1845 to 1919 when the mine closed early in that year because of a prolonged miners' strike.

Underground working ceased at Foxdale in 1911. Heavy pumping and mining costs led to the closure of the mine when all work at higher levels ceased to be remunerative. Foxdale from official records from 1845 produced 143,756 tons of lead ore yielding 2,882,440 ounces of silver to 1900. Snaefell mine continued operating until July 1908 when a fall of rock occurred in the shaft wrecking and blocking it from 40 fathoms downwards. The mine had been unprofitable for a long time so there was no inducement to reopen it. The famous Bradda vein was worked intermittently and fitfully and was for long periods unworked. The ore was carried by boat to the smelter at Port-St-Mary.

Lead mining in the Lake District goes back to the twelfth century and became prominent in the great revival of metal mining in Tudor times. The Caldbeck Fells, the Keswick and the Helvellyn fields were the three lead mining areas there.

The most important of the Caldbeck Fells mines was the ancient mine of Roughtongill seven miles from Keswick at the head of Dale Beck. It was worked in the twelfth century but was abandoned when the Germans took charge in 1566. In 1571 of the eighty employees twenty-three were German miners. The most important of the four veins was the Great South Lode. Lord Wharton successfully worked the Silver Gill section from 1710.[21] A smelter was built there in 1794 which was still functioning in 1843, producing 100 tons of lead a month. Eriggith mine was on the same lode but was not developed until comparatively recent times. It was first mentioned in 1790. The outcrop was rich and the quantity of ore raised justified the erection of a smelter by Carrock Beck. In 1810 new owners erected an improved dressing plant which in 1813 was working night and day. This company worked the mine until 1822 producing 200 tons of lead a year with a silver content up to 30 ounces. William Jefferey reorganized the mining and dressing plant and up to 1857 nearly two thousand tons of lead and copper concentrates were produced.

The Keswick area contains many old mines producing lead ore including the historic Goldscope Mine, but its reopening was not successful.

The Helvellyn field covers forty square miles of rugged mountain country. Though lead mining here goes back to Tudor times, it was only practised on a modest scale until the nineteenth century. The most important mine was Greenside which operated almost continuously for 150 years to 1962. This mine lies on the steep mountain slope west of Ullswater and is in Westmorland. The vein was probably discovered in the mid-seventeenth century and in 1690 a party of German adventurers drove the top and middle adits sending the dressed lead ore by pack-horse to a smelter near Keswick, but by about 1800 the London Lead Co. had erected an up-to-date smelter at Alston to which the dressed ore was carted. The Greenside Mining Co. was formed in 1822 and mining started before 1825 with an output of 1,500 tons of dressed lead ore a year. The company depended on dams and water power as it used hydraulic engines. During the 1830s a smelter at the foot of Lucy Tongue Gill was built with the flue along the hillside a mile long which, with the stack, gave a 1,000 feet draught. This stone-arched flue followed the bedrock and formed a wonderful crosscut covering every possible extension of the Greenside vein. William Henry Borlase became manager in 1890 and persuaded the company to install a hydroelectric plant for pumping, winding and operating the new compressor. The winding engine at Smith's shaft was the first electric winding engine in a British metal mine and the electric locomotive replacing the ponies and drivers was also a first and gave good service for forty years. By World War I smelting became uneconomic and the concentrates in wagons drawn by a steam tractor went to Troutbeck station and then by rail to Walkers Parkers Newcastle smelter.

During World War I most of the young miners were called up, and little development work was done. Access to the mine was by the Lucy Level and internal shafts. Efforts after the war to catch up took the workings down to 135 fathoms below the Lucy Level by 1925. A cloudburst on 29 October 1927 wrecked part of the Kepplecove Dam embankment, damaging shops and houses in Glenridding at the bottom of the mountain. This disaster and low metal prices combined to stop mining in 1934, but maintenance was continued. The Basinghall Mining Syndicate then took over in 1936, erected a modern 250 tons a day grinding and flotation plant. Luckily they were able to take power from the extended National Grid and operated successfully until 1959. In 1960 the U.K. Atomic Energy Authority carried out explosion tests underground. The last ore reserves were gleaned in 1961 and with all ore exhausted the mine closed for good in 1962. Other mines in the Helvellyn area were Hartsop mine, small but silver-rich with 30 ounces to the ton of ore, and with its old entrance in a wood.

Lead ore in North Wales has been worked more or less continuously since before Roman times. At Ffos-y-bleiddiaid Mine, or Wolves Fosse, near Abergele, curious hammers and other tools were found and also the golden hilt of a Roman sword. There is a tradition that a large town stood at Croes-Ati near Flint as the plough has often turned up the foundations of buildings and quantities of lead slag, lead ore and bits of lead metal just above the shore in Northop parish. Tons of lead have been got within a short time at Pentre Ffwrndan—the place of the fiery furnace.[22] Large quantities of Roman coins, brooches and buckles and other artefacts prove that Flint was the Roman port for exporting lead. The discovery of several Roman lead pigs near the borders of North Wales makes clear that the ore was smelted in this region.

In 1283 Edward I granted the burgesses of Flint a licence to cut timber from the woods of Northop and neighbouring parishes to smelt their lead ore and Charles I granted the working of all the lead mines within the hundreds of Coleshill and Rhuddlan to Sir Richard Grosvenor, the ancestor of the Dukes of Westminster.

In 1728 a rich find of lead ore was made near the Pant Lode in the Halkyn district. Chester exported 1,000 tons of lead metal and 300 tons of lead ore in 1771 and 3,470 tons of metal and 431 tons of ore were sent by sea to London and Liverpool for re-exportation soon after. The village of Pentre Halkyn was created by a discovery of a rich vein of lead between 1860 and 1870 and Holywell town grew about the same time for the same reason. Fabulous amounts of lead and zinc ore were also got from mines in the Llanarmon district and the famous Minera mines were worked intermittently from at least Roman times. They were flourishing in 1766 when the day level was first considered to drain the mine. Smelters were established from time to time at Bagillt, Connah's Quay, Flint and at Hawarden, the home of the Gladstone family.

North of the Bala Fault, the lead and zinc ores are restricted to the carboniferrous limestone and the overlying sandstone series with the exception of Bryncelyn mine near Rhydymwyn where ore was won from the Lower Coal Measures. The veins run in an east to west direction whereas the crosscourses run north to south. These crosscourses contain no blende, are poor in silver, while copper pyrites occur only as small specks. The galena also is more earthy than in the veins. They are often barren with large empty caverns and have formed the chief underground water channels for many million years. The flats were more like pipes than the Derbyshire flats as they were irregular beds in a mass of calcspar. When blende and galena occur together within an ore shoot, the blende increases in depth. Talargoch, Talacre and Minera are examples of this feature but the distribution of silver does not seem to follow any rule and varied from 0 to 18 ounces a ton of galena. Recorded observation suggests that in some lodes the silver content increased with depth, a phenomenon observed also in mid-Wales.

The greatest deposition of ore occurred in three stretches of country: Prestatyn to Holywell, Holywell to Llanarmon and, south of the Bala Fault, at Minera. Ore only occurs sparingly on the west side of the

Vale of Clwyd near Abergele, at St Asaph, and between Minera and Llangollen.

Water was the obstacle to deeper mining and this whole area is blocked out by veins and crosscourses into a chessboard pattern dipping obliquely seaward and resulting in different water-table levels in contiguous blocks which were upset only when pierced by tunnelling operations. The mining in both counties depended on deep drainage.

The Halkyn tunnel was commenced in 1818 and started from a stream at Bryn-Moel, 200 feet above O.D. A company formed to extend the tunnel in 1875 drove it a further four miles. When this drainage company was formed, the mines were almost derelict and it successfully dewatered, wholly or in part, mines as far south as Llyn-y-Pandy.[23] The driving of the Halkyn tunnel allowed mine by mine to be reopened and after this tunnel was completed the Flintshire mines contributed 20 per cent of the country's lead metal.

The Milwr sea level tunnel was undertaken by the Holywell-Halkyn Mining and Tunnel Company in 1896. It began in July 1897 at Dee Bank, Bagillt, on the Dee estuary and was driven three and a half miles. By 1908 the face had reached a point 600 yards south of Herward shaft. In 1913 the Halkyn District Mines Drainage Co. secured another Act of Parliament to extend the tunnel into the Halkyn Mines area in order to drain the mines to a further depth of 190 feet. Windmill was reached in August 1919 when work stopped. The famous St Winifred's Well, some three miles north by west of the Milwr tunnel's heading, dried up soon after it was completed. On 5 January 1917 the tunnellers broke into a large sand-filled loch on the Pant lode and the whole watertable north of this lode was lowered. Because of its religious significance, the drying up of St Winifred's Well caused consternation, but a combined pumping and drainage scheme restored the flow to the well and to Holywell town as well.

At Minera it was proposed to drive a tunnel from the River Alyn to the Minera mines to secure 300 feet deeper drainage, but the cost was estimated to be fully £100,000 so the scheme was never carried out though many thousands of tons of ore are still there under water.

On the east side of the Vale of Clwyd at the northern end near Dyserth, Talargoch mine was most prominent.This was a Roman mine and the outcrop down the rocky sides of Graig-Fawr would have attracted immediate attention. In earlier days a great deal of gravel ore, masses yielding up to 80 tons of galena, were found in the base of the glacial drift. From 1845 to 1884, 57,752 tons of galena were produced but in May 1884 pumping stopped and the mine filled with water to day level. After that there were only gleaning operations on the extensive dumps. The Talargoch vein runs the whole length of the mine, about 1400 yards. Production before 1845, though considerable, is not recorded.

Talacre mine is south of Gronant near the coast. Up to 1884 it was worked near the fault that throws the chert against the Lower Coal Measures and the workings were all in the 105 yards thickness of chert. Like Talargoch, it yielded gravel ore and it also produced calamine which had been worked at the mines near Holywell since 1740 and was found near the outcrops. It was called 'coke'–not to be confused with 'cawk', the name for barytes. Up to 1861 the mine produced more lead ore than zinc ore but from 1870 to 1884 zinc ore replaced the lead ore in depth. The Old Holloway mine was drained by an adit that opened into the Holywell valley only 200 yards north-west of St Winifred's Well, about 230 feet O.D. This adit was begun in 1774 and was called the Boat Level as it was used as a canal to transport the ore in barges. Thomas Pennant described a voyage he made, accompanied by his two sons, on 21 September 1795. He mentions the rocks they passed and the cascade of water 1,167 yards from the mouth of the level. From 1877 the mine was managed by a Captain Hotchkiss for ten years. A number of veins were cut and, at the True Blue lode, water issued from both sides, which was called 'Harrogate' water presumably because it smelt of sulphuretted hydrogen. The Merlyn vein was worked to provide ore worth £150,000, raised from a depth of only 20 yards in 1850.

At the Milwr mine this great vein ranged for a total distance of three and a half miles. The richest ore was at the eastern end. The system of veins and crosscourses was drained by a day level begun in 1754 which emptied on to the south bank of the Nant stream.

In the Picton mine a water-filled cavern was cut with so little warning that the miners barely escaped with their lives. The water soon disappeared but temporarily discoloured St Winifred's Well, and an actual explosion occurred in a rise above the Blende Sump. The ore was taken out of the mine in tubs drawn by ponies to dressing floors at the tunnel mouth. One pony hauled out 25 tons of ore a day.

There is a maze of east-west veins and north-south crosscourses on Halkyn mountain on both sides of the Nant Figillt Fault. North of the fault the principal producer was the Halkyn Mining Co. Ltd., reputedly started by Captain Matthew Francis with a capital of £10,000. He and his son Jack are mentioned several times in the special report of this area in the *Memoirs* of the Geological Survey. They were son and grandson respectively of the Absalom Francis mentioned below in describing the Mid-Wales lead mines.

There were two main lodes in these mines, the Pant-y-Gof and the Great Halkyn veins. They were dewatered by the Halkyn tunnel. The maximum thickness of the vein was from 20 to 30 feet with 6 to 9 feet of galena. Up to 1920 the output was 40 tons of galena and 50 tons of blende a month. The all gravity concentration plant was near Lewis's shaft. Flotation was mooted but never adopted. The coarse galena went to the potteries for glazing, mixed grades from both lodes went to Murex and the galena to Walkers Parkers' smelter at Bagillt. Powell's lode, discovered in 1903, was connected to the Great Halkyn lode later.

In a field near Gelli-Fowler, at the intersection of a crosscourse and a vein, two men raised £40,000 worth of ore in a few years. Pennant wrote 'The richest vein was discovered (about 1728) at Rowley rake . . . which in less than thirty years yielded to different proprietors, adventurers and smelters above a million of money.'[24]

In the nearby Prince Patrick mine, a band of galena looked like a coal seam 2 feet thick. The Union vein yielded a great deal of calamine and Long Rake, with most of its course in limestone, was worked opencast since Roman times. A remarkable occurrence in both the Old Rake and Pant-y-Ffrith crosscourse was the discovery of tree trunks at great depths, presumably due to the surface being engulfed by the collapse of the limestone. At the California vein of the Central Halkyn Mining Co. a lot of lump ore in clay was found at a depth of 156 yards. It was exported to West Africa to be powdered for eye paint.

On the south-west side of the Nant Figillt Fault there was just as great a chessboard pattern of veins and crosscourses in an escarpment of limestone that included the Hendre and Mold mines. At Bryn-Gwiog mine in 1860 the lode of galena and blende was 4 feet wide but had a highly detrimental fluorspar content. The Hendre area produced a great deal of lead ore up to 1880, the yield being mainly from pipes hollowed out from the limestone. One flat occurred as a nearly horizontal pocket under a shale roof and, though not the largest, it contained 2000 tons of galena. The Bryn-Celyn vein runs from the river at Nant Alyn to Rhydmwyn Foundry. It was dewatered by the Halkyn tunnel about 1893 after having ceased work in 1845 due to water difficulties.

The extension of the Halkyn tunnel met the Llyn-y-Pandy lode in 1901 and drained its western end.

Below is an account by the Rev Richard Warner of a visit he made to this mine in 1798.

> 'Another hour brought us to the great object of our day's ramble, Llyn-y-Pandy mine, the most considerable lead mining speculation in England. . . . Llyn-y-Pandy mine is the property of John Wilkinson Esq., the great ironmaster, who has . . . brought it to its present state. . . . With all his exertions however, he has not been able to render it complete. The mine . . . contains so much water he has been under the necessity of erecting four vast engines of Boulton and Watt's . . . to drain it. . . . Many thousands of tons of lead ore are now in stock . . . waiting for a market, the war having almost suspended the demand for lead. The engines also are quiet and the works at a stand. The bottom level . . . contains one head of solid ore upwards of 6 feet wide. A little down the river . . . is a lead mine called Pen-y-Fron (Bryn-CelyN) belonging to Mr Ingleby-drained by a steam engine and a waterwheel . . . Mr Ingleby is scarcely ever able to get to the bottom of his work . . . were he able to effect this completely his profits would be immense since the mine is incalculably rich there being one vein of solid ore two yards and a half in width. . . . Smelting houses are in the immediate neighbourhood of the river . . . and about half a mile below . . . is a mill . . . for rolling the lead into sheets.'[25]

After this time little was done at Llyn-y-Pandy. In 1828 an 80-inch Cornish Engine was erected but the water proved to be too much for it and the work was abandoned. This engine went to Cat Hole and later to Minera where it gave half a century of good service.

The Pant-y-Mwyn vein extended nearly the whole breadth of the limestone, the Cefn-y-Fedr sandstone and even for some distance into the coal measures. It was exploited from 1823 until 1844, when a large run of water at 238 feet in Taylor's Shaft stopped all work. It was reopened in 1901 and was taken over in 1910 by Brunner Mond & Co., who got a fair amount of blende and some galena, the blende being satisfactory for the Brunner-Mond zinc process. This company was active in these southern mines and, between 1890 and 1903, carried out a considerable amount of work at the Gwyn-y-Mynydd and Cat Hole mines. Cat Hole, Gwyn-y-Mynydd, Westminster, Bog and Belgraev mines were all rich and prosperous up to about the middle of the nineteenth century. A former agent of the Duke of Westminster stated that about £3 million worth of ore had been got from the Pant Lode and the Bog mines. At Cat Hole mine, tree trunks were found deep underground and their presence, like those at Pant-y-Ffrith, was presumably caused by the collapse of the limestone into a fissure.

The story of lead mining in Denbighshire is that of the Minera mines. Mining was active there from before Roman times up to the beginning of World War I. The limestone dips generally south-east and on Minera Mountain is succeeded normally by the Cefn-y-Pedw sandstone. Near Minera the strata are thrown down by a complex belt of faults. The most important veins were the Red Vein, which stretches from the Bala Fault to the Wern and the Main Vein, which continues for two miles passing into the Coal measures. Ore was got from it as far south as the Nant Minera. These two veins are more or less parallel for three-quarters of a mile northwards when the Main Vein veers westward. In 1865 a mining engineer named Brendon Symon mapped the older mines, but there are no old plans deposited with the Home Office.

At the north end of the system practically all the faults and joints were ore-bearing. The earliest mines were apparently on high ground on the north side of the Clywedog valley from which in bygone days adits were driven into workings with extensive caverns, one 428 yards long. The extent of these old workings was remarkable.

In 1815 water from the west mines found its way through an open fissure to the deeper mines in the east, making further work impossible for some years. Pumping was organized in 1817 when it was planned to raise 4,000 gallons a minute, but there were so many breakdowns of machinery that the work was abandoned in 1823.

The Minera Mining Co., formed in 1845, extended the deep day level at 830 ft O.D. and about 1849 connected the east and west mines and dewatered the whole minefield some 80 yards lower than before. The deep day level leaves the vein near the Wern and issues as a 3 feet wide, 1 foot deep stream near New Nant Mills. The richest bunch of galena was found in 1864 on the Red Vein between Ellerton and Lloyd shafts 6 to 7 feet wide at the 200 yards level under a shale roof and it realized £500,000 when worked by the miners for 11*s* a ton on tribute. In 1909, when pumping finally ceased, there were three large pumping engines at three separate shafts, costing £600 a month. It took eighteen months for the mine to fill up to day level. The company was wound up at the beginning of World War I and during the few intervening years ore was profitably won from above water level.

Harking back to Halkyn, the sea level tunnel from Milwr was driven to Windmill in 1919. A new company, Halkyn District United Mines, was formed during the 1920s. Its first step was the amalgamation of all the local mineral rights involved, some large and numerous small owners. After long and wearisome negotiations by the deputy-chairman, Noel Humphries, this was eventually accomplished by 1928. The area so consolidated covered twenty-five square miles. The first step was to extend the 1919 tunnel face to Penybryn shaft, a distance of half a mile. This was accomplished without the completely modern and carefully researched procedure used from Penybryn. In fourteen weeks to 5 July 1930 the advance made in this 10 feet wide by 8 feet high tunnel was 2,037 feet which was claimed as a European tunnelling record (see Plate 21).[26] In due course this tunnel cut lodes and became a producing mine. In the early 1930s, from an advance of 6,650 feet of tunnel a quarter of a million tons of ore were produced. Unfortunately, however, as fluorescein tests proved, the River Alyn was leaking into the mine through numerous swallow holes. At great expense an attempt was made to seal them but this was completely ineffectual and on more than one occasion the tunnel became so seriously flooded that transport became impossible and all miners had to be withdrawn as the water on the raised underground railway was up to waist height (see Plate 17).[27]

The area covered by North Cardiganshire and West Montgomeryshire is the wildest part of mid-Wales. Up to the beginning of World War II there were isolated farms in narrow but fertile valleys between the hills where no word of English was spoken or understood. Around Plynlimon Fawr is a desolate region where the rivers Severn, Wye, Ystwyth and Rheidol have their source, the last two meandering westward to empty themselves into Cardigan Bay at Aberystwyth. The rock formations of the region are Silurian and Ordovician, the former comprising the Wenlock and Cwm-Ystwyth and Frongoch formations and the latter those of the Gwestyn and Van. The region forms a compact unit throughout, in which the character of the rocks, the lodes and their fillings, is fairly uniform. Of the 128 mines listed some of the most productive were worked by John Taylor & Sons for the best part of half a century up to the 1880s. One of their mine captains, Absalom Francis, mapped the veins in most of the area and, though academically unqualified, was a geologist of some renown.[28] A few of the most interesting mines will be briefly described.

Esgairhir and Esgairfraith mines acquired some renown from reports published by Waller between 1698 and 1700. Waller claimed that with 600 men the mines would yield a yearly profit from silver lead ore alone of £70,000, compared them with Potosi and described the cluster of newly erected miners' houses as the 'new town called Welsh Potosi'. In 1693 Sir Carberry Price procured an Act of Parliament in the fifth year of William and Mary whereby the Crown relinquished its claim to mines notwithstanding they contained gold and silver. Esgairhir was worked spasmodically until 1744 and then remained unmentioned before being actively working again in 1810.

Esgairhir is drained by an adit which intersects the lode at about 135 fathoms. From there it was driven on the main lode and a branch opens on to a marshy tract near the miners' houses. So it was open at both ends and served as a road for the miners going into and leaving the mine. From the surface to the adit no fewer than eighteen shafts were sunk. The ore occurred in pipes. A large amount of ore was won before official returns were made in 1845, possibly altogether 10,000 to 20,000 tons. These two mines lay on the north-west edge of Plynlimon Fawr.

Alltycrib Talybont is one of the oldest mines in the area and is about half a mile from Talybont. The ground was drained by a deep adit about 170 O.D. It was said that large quantities of Potter's ore were obtained above an adit called Wilkin's level 250 O.D. Bronfloyd (a corruption of Bryn Llwyd) mine, one of the oldest mines, was worked by Thomas Bushell about 1645. He purchased it from the widow of Sir Hugh Myddelton. The mine was producing at intervals for some 250 years prior to the mid-nineteenth century. It is one of the most westerly of the mines, being two and a half miles east of Bow Street station.

East Daren or Cwmsymlog mine was worked intermittently for several centuries before 1901 when it closed. The earliest record was when it was worked by the Society of Mines Royal. When taken over by Sir Hugh Myddelton the profits he made there enabled him to undertake the New River projects to bring water from Ware to London. Sir Hugh is stated to have made a profit of £2,000 a month out of working this mine. Afterwards Thomas Bushell, Sir Humphrey Mackworth and the company of Mine Adventurers took over and worked it profitably and continuously from 1750 to 1770. Messrs John Taylor & Sons acquired the mine in 1847 and carried out a good deal of exploratory work at the western end, while numerous trials were made at the eastern end.

South Daren or Cwmsebon mine lies in the Erfin valley and is about five miles from Bow Street station. It is the deepest mine in the whole area and was only developed since 1840 when in sinking the engine shaft a rich shoot of ore was discovered. A new shaft was sunk to the 154 fathom level, a total depth from the surface of 166 fathoms or just under a thousand feet. In spite of its depth it was a comparatively dry mine.

Great Daren mine is half a mile west of South Daren and was worked for a long period. Sir John Pettus referred to it as a Roman work but there is little evidence to connect the Romans with any lead mining in this part of Great Britain. The Mines Royal took over Great Daren under a patent granted in 1568. Here there is an oval hill camp probably Ancient British, and, along the outcrop, a wide trench 600 yards long and still in places as deep as 60 feet. The mine was in its prime from 1720 to 1740 and was reopened about 1825. It then worked intermittently until 1879. However, the output since 1845 gave little indication of earlier riches.

Cwmerfin mine lies in the Erfin valley about five and a half miles from Bow Street station. There is some confusion both as to the age of the mine and whether it is one mentioned by Pettus as worked by the Society of Mines Royal, but it definitely was worked successfully by John Taylor & Sons from about 1848.

Goginan mine lies about half a mile north of the Llanidloes-Aberystwyth road and about seven and a half miles from the latter town. This mine was worked by the Society of Mines Royal and Waller published a section of the work in progress in 1699. The lower adit was cut by chisels. Taylors took over in 1836. The richest deposits were above the adit or 60-fathom level but it was still productive down to the 120-fathom level and it produced nearly half a million ounces of silver.

Old Esgairlle and Great West Van mines are also near the main Llanidloes to Aberystwyth road and about six miles from Devil's Bridge station. The large dumps prove that the work was on a considerable scale.

Frongoch mine is about two miles SSW of Devil's Bridge. It was on the property of the Earl of Lisburne and was discovered by J. Probert of Shrewsbury in 1798. Three years later Probert discovered Llwynwnwch mine immediately east of Frongoch. These properties, both known as Frongoch, were taken over by Taylors in 1834 when the deepest working was only 34 fathoms below surface. At that time the output was less than 40 tons a month. In the first year under Taylors it rose to 100 tons a month. In 1879 Taylors sold the mine and dressing plant to the Earl of Lisburne. It was then taken over by John Kitto who broke down the walls of the stopes where blende had been left standing and in one year won 2,850 tons of blende. When Kitto left a Belgian company took over but it did little work underground.[29] The mine has been idle since 1903. It was abandoned by Taylors in 1879 for a number of reasons; rock hard to break and a large amount of water to pump by waterwheels. Owing to frost in winter and drought in summer there were too many periods when no pumping could be done and therefore no work was possible in the bottom levels. A steam engine meant coal and transport for four and a half miles from Crosswood G.W.R. station. So, although in the three years 1875-78, 3,205 tons of lead ore sold for £44,857, there was a net loss of £679.

Cwmystwyth mine is three miles from Devil's Bridge on the north side of the River Ystwyth and the workings were in three sections, western, central, and eastern. There were signs of prehistoric workings and this mine was exploited for several centuries fairly continuously. The western workings carried galena not usually mixed with blende or pyrite. The central workings carried galena and blende with no pyrite, the eastern workings a mixture of blende, galena and copper pyrites in a very wide open lode. From 1848 to 1916 this important mine produced 32,912 tons of lead ore.

Logaulas mine has an odd history. An adit was begun in 1785 but was still not completed in 1810. When the lode was eventually cut it was barren and the mine abandoned. Afterwards, some Cornish adventurers drove the adit a few feet further and cut the true lode with a vast deposit of ore that gave them for several years a rich reward. They then lost the true lode, mistaking a small vein for it to the south which was so poor that they quit dispirited. John Taylor & Sons then made an accurate survey, drove a crosscut to the north and cut the lode, finding rich ore only a few feet from the scene of their predecessors' failure. Taylors took over Logaulas mine in May 1834 and worked it profitably for forty years. The horizontal extent of the lode was 3,600 feet but was surprisingly shallow, being limited to the 120-fathom level and much of it only down to the 90-fathom level.

Esgairmwyn was discovered about 1751 under the superintendence of Lewis Morris who was then steward for the Crown. The mine had the second deepest shaft in the district. This was inclined and extended to 155 fathoms below the surface. In driving southward a large volume of water was tapped which drowned a part of the mine. From 1852 to 1917, 10,430 tons of lead ore was won. An odd anomaly was that the silver content improved in depth from under 2 ounces to over 16 ounces to the ton of ore.

Van mine was the biggest in this whole district. It lay about two and a half miles NNW of Llanidloes and was served by a branch railway line. Previous to 1850 a search was made for the extension of the lode eastward and after four years the lode was found near a large quarry. From this point an adit was driven eastward, but it was eight more years before ore was found in a winze below the level. Owing to water difficulties and the small amount of the ore, the manager decided to drive on a crossent begun some years before and, at 75 fathoms, at last struck a rich lode about 30 fathoms below surface. This was fifteen years after prospecting started. A company was floated with 12,000 shares of £4 5*s* and in less than five years these were quoted at £84. The ore from all the workings was drawn along the 120-fathom level to the engine shaft and through the adit to the dressing floors. The cut and fill method of stoping was adopted. The ore was 40 to 50 fathoms wide but only 20 or 30 feet deep. The lead ore produced was 96,739 tons and the silver contained was 771,557 ounces.

At Snowbrook there was an ancient mine where, when it was explored in an unsuccessful attempt at reworking in 1859, small picks and wedges were found, a broken stag horn and a smooth, round, 3½-inch diameter stone ball. There are extensive traces of old shallow workings. Probably this was a prehistoric Ancient British working.

In the north corner of Montgomeryshire there is Llangynog mine where lead ore was discovered in 1692 in a lode up to 3½-yards wide of which 2 yards was solid ore. It was early mined to a depth of 300 feet when water stopped further working, but it had been flourishing for forty years. In 1881 D.C. Davies said that the mine had been worked for 150 years. About 1797 it was being worked for calamine as well as lead, the products being sent to foundries near Ruabon. In 1809 it was described by Westgarth Forster of the Beaumont company as the richest lead vein in Great Britain. It was revived in 1845 when sixty men were employed there and from 1852 to 1880 it worked under the name of the Chirk Castle Mine.[30]

In Merionethshire, the Bwlch-y-Plwn mine, three miles from Penrhyndeudraeth station, was extensive and reputedly produced large quantities of lead ore from seven lodes. The outcrop of the Gwyn mine, eight miles from Dolgelley, can be traced for a mile and a half, galena and blende occurring at a depth of 120 feet, but it was notable as the principal producer of Welsh gold for many years. From 1864 to 1907, 36,116 ounces of gold were won from less than 100,000 tons of rock. It was said to have considerable reserves of lead ore.

In the Llanengan area there were seven mines on the St Tudwals peninsula but all had been abandoned by the beginning of the twentieth century. Collectively they produced, from 1873 to 1892, 19,965 tons of dressed lead ore from which 13,520 tons of metal was obtained.

In the Llanrwst district remains prove that the Romans explored the area. John Williams observed that Thalaspi Alpestre (Alpine cress) grew abundantly, especially near the lead mines, an observation said to be verified in other countries. This area, from near Bettws-y-Coed to Trefriw, formed the principal lead-zinc field of the mountains of North Wales. Parc mine is a mile south-west of Llanrwst. Here a pipe of lead ore yielded 16,000 tons of galena. The adit level in 1919 was 1,850 feet long. The lodes in the Llanrwst area consist of a breccia of slate or ash fragments cemented by calcite, galena and blende. This district is bounded by the Conway Valley on the east and a ridge of hills about three and a half miles to the west. Attempts were made in recent years to revive mining at Parc mine. The Mitchell brothers of Wanlockhead were active there and after World War II an international mining company made an unsuccessful attempt to bring it into profitable operation.

The Gorlan mine in 1882 produced 621 tons of ore which on smelting yielded 2,233 ounces of silver.

The Trecastell mine is three miles south-west of Conway. The two main lodes were different from the Llanrwst group in that the blende and galena lay adjacent to the ash without any breccia of country rock or spar. There was no water problem and all that was pumped from the sump was used in the dressing plant. Said to have been worked by the Romans, this mine was the principal producer of lead and zinc ores in this part of North Wales. From 1892 to 1913 it produced 6,948 tons of lead concentrate.

An odd occurrence of lead/zinc ore occurs in the bluestone of the Mona Parys copper mines near Amlwch in Anglesey. Bluestone was the principal ore of the Black Rock and the Morfa-Du lodes and is a dark, bluish-grey material, heavy and of finely grained texture. From the Home Office Statistics, the lead and silver obtained from this Mona mine bluestone between 1882 and 1911 was 1,262 tons of lead and 84,426 ounces of silver.

West Shropshire was one of the several lead mining areas fully exploited by the Romans, who evidently got important supplies of lead ore from outcrops and shallow underground workings. That the Roman vein in the Roman Gravels mine was worked to more than 300 feet in depth by the Romans was proved by the discovery of mining tools in the ancient galleries. It is one of the largest veins in the district and was worked for a length of half a mile. In the old workings on this vein curious wooden shovels and candles with hempen wicks, coins, pottery and a pig of lead were found. The pig was marked IMP ADRIANI AUG. Two other pigs of the Emperor Hadrian (A.D. 117-138) have been found in the district.

The Roman excavations were enlarged during the twelfth and thirteenth centuries when lead in abundance was won in this area owing to the then high demand for lead for roofs, cisterns, pipes, and other essentials in the building of castles and monasteries. Unfortunately there is no record of the output for such early periods.

The mines lie between a mile and a quarter and five miles south and a little west of south of the town of Minsterley. The most productive in recent times was Snailbeach, about a mile and a quarter south of Minsterley. There are plans of this mine in existence dated 1790 and in opening up the mine for barytes many old workings were encountered. The vein trends mostly east to west and was worked to a depth of over 552 yards from numerous levels 30 yards or more apart. The ore in the western half of the mine was more patchy. The greatest amount of ore was won from below the 192 yard level in the eastern half. The main lead-bearing vein had regular walls and at times was as wide as 22 feet. In the richest parts, the galena was in shoots 120 to 150 feet long. Zinc blende increased in depth. There were large cavities in the hanging wall one such vugh was 48 feet long, 21 feet high, and a foot wide. The recorded output from 1845, when official records started, to 1909, when the pumps were stopped, approximated to 130,000 tons of lead ore. Before pumping ceased, twenty-one men in five weeks won 65.6 tons of lead ore between the 492 and 500 yard levels.

East Roman Gravels and Roman Gravels Mines are a quarter of a mile apart and the latter a similar distance north of the village of Hope.

It was in the latter mine that many ancient artefacts of Roman date were found. Here the vein was worked opencast and in 1839 the mine was being worked at a depth of 336 feet below the ancient Roman galleries.

The Grit mines were an important Shropshire group. They lie about three-quarters of a mile south-west of the village of Shelve, where there are three important veins called the River, the New and the South. The first has been stoeppd from outcrop to below the 60-fathom level. Where the New vein crossed it there was a large swallow at the 60-fathom level into which the ground collapsed, causing the rich ground to be abandoned. This swallow extended downwards for an unknown distance. In the West (or White) Grit Mine, the veins are the Rider and the Squilver, which is a corruption of 'disgwylfa,' a Welsh word meaning a place of observation.

The Boat level, whose outlet is about 900 O.D., drains the Tankerville and Pennerley mines and then turns SSE to Bog Mine. The total length of this drainage adit and its branches is about two and a half miles. Tankerville or Oven-Pipe mine is so named because of the peculiar shape of its ore-shoots which are like fingers following the 50 to 60 degrees dip and of about equal width and thickness. These bodies are a combination of vein and flat and have been worked to a depth of 90 fathoms below the Boat level, or from the surface 1,050 feet. The mine has long since been abandoned. All the more southerly mines were drained by the Boat level. An attempt to dewater the flooded mines by a deep adit was made by Shropshire Mines Ltd whose moving spirit for many years was Col. Ramsden. This adit, which was known as the Leigh level, had been started by the Farmers Company about 1820 with a capital of £39,000, the intention being to drive a deep tunnel 7 feet high by 5 feet wide which would drain all the Shelve mines. However, owing to money spent on lawsuits with Lord Tankerville, it was abandoned in 1835 for lack of capital after being driven for a mile and a quarter. At the beginning of the 1920s, the Shropshire Mines Ltd determined to extend the tunnel, which is about 397 feet O.D., and thus drain an immense area to from 800 to 1300 feet below the surface. The expectation that the tunnel would, during driving, open up productive ground was not fulfilled. Since official records were started in 1845 West Shropshire mines produced over 200,000 tons of lead ore. The highest recorded output of lead concentrates in this district was between 1871 and 1875 when it reached in one year 8,000 tons. After that date there was a rapid decline.

One authority is of the opinion that lead from the Mendips was produced centuries before the Roman invasion and possibly transported to Cornwall and exported along with Cornish tin. In the British Museum is a silver-lead pig from Charterhouse stamped with the name of the Second Agusta Legion and dated A.D. 49. Another similar pig was found at St Valery-sur-Somme in northern France. Four more pigs stamped for the Emperor Vespasian were found twenty years ago at the point where the Wells to Bath road intersects the Roman road from Charterhouse. At Charterhouse, archaeologists excavated a large mining settlement including a small amphitheatre of undoubted Roman age. The mining community there seems to have been housed tolerably comfortably. There is no doubt that the Romans were exploiting this silver-lead mining area of Somerset only six years after the invasion. They smelted the ore on the spot, leaving behind them large heaps of debris consisting of ore and slag.

We know that in medieval times the Mendip mines were active, because in 1184 Richard I granted the Bishop of Bath the right to work all the lead ore on his land in Somerset, and Henry III conferred on a later Bishop of Bath the right to mine for lead and iron ore, on terms, in the Forest of Mendip; also the right to cut timber in 40 acres of woodland in Cheddar as fuel to smelt the ore. The earliest customs of the Mendips were written down in the form of agreed rules, and during the reign of Edward IV (1461-83) there was a great debate over the conflicting rights of the commoners and miners. The king sent the Lord Chief Justice of England who, 'sitting at Lord Bath's place heard the four Lords Royall of Meynedeepe and some ten thousand other persons to seek agreement on the regulation of the mines'.

In the opening words of the proclamation the Old Mendip Mining Laws declared:

> 'If any man whoever he be intends to venture his life to be a workman on the said occupation of a miner; if any man of this doubtful and dangerous occupation does by misfortune take his death by the falling of the earth upon him by drawing or stifiing or otherwise as in times past may have been, the workmen of this occupation are bound to fetch the body or bodies out of the earth and to bring him or them to surface to a christian burial at their own proper cost and charges although he be three score fathoms under the earth.'

This without interference of the coroner.

These Mendip laws were enforced up to the early seventeenth century, and additions printed in 1687 included 106 separate items. They were engraved upon the side of a map, the Chewton map of the Mendip Mines, prepared between 1461 and 1485, which is now preserved in the Waldegrave Estate Office at Chewton Mendip. The old medieval custom was that a man must first get a licence from the Lord of the Soil, a privilege that could not be denied him. He then from his trench threw his hack (a pick or hoe) both ways and as far as he threw no other man could work therein. Theft of lead metal or ore led to the destruction of the culprit's tools and his banishment from the mines. After the miner had washed his ore and it had been smelted, one tenth of the metal went to the Lord of the Soil. Fire-setting was still being practised in 1668.[31]

Thomas Bushell, who has been mentioned previously, had a scheme for dewatering the Mendip mines by a great adit but this did not gain the support either of the Lords of the Soil, or of the miners, as neither was willing to bear the cost of so ambitious a scheme. Nevertheless the lead output of the Mendips reached a maximum of about 4,000 tons of metal a year in the first half of the seventeenth century. The metal appears to have been arsenical[32] and thus admirable for the making of lead shot at the Bristol shot tower, the first such tower in the country. There is an interesting account of the dressing and smelting of the Mendip lead ore published in 1668, stressing the evil effect of lead dust and fumes on man and beast.[33]

In 1856, the year of maximum output of lead in Britain, the Mendips were credited with 750 tons of material for smelting which produced 500 tons of lead metal. However, this was *not* freshly mined ore but lead slag and galena gleaned from the huge heaps of debris. There were three small firms extracting lead from this source in 1870: Treffrey and Co.'s Mendip Hills Lead Co., East Harptree Lead Work Co. Ltd, and St Cuthbert's Lead Smelting Co. Ltd, near Wells. None of these three bought lead ore on the open market. They were scavengers.

The Mendip Mines, like other lead mines in limestone country, suffered greatly from incoming water and from extensive caverniation—hence Bushell's bold but unadopted scheme. In August 1967 four local students exploring the underground abandoned workings on Mendip discovered a cavern at Priddy on the 900 foot contour 2,500 feet long and 80 feet high.

The ores of the lead mines of Devon and Cornwall were remarkable for the richness of the silver they contained.[34] An estimate of West Country lead metal output was half a million tons, or double that produced during the few years from 1845 to 1886 when records were kept, for between these two dates, according to *Mineral Statistics* compiled by Robert Hunt, the two counties produced 311,049 tons of lead ore between them.

Many small lead mines have been worked periodically near Okehampton and Tavistock in west Devon, none of much importance except Wheal Betsy (North Wheal Friendship) in the parish of Mary Tavy. This was a very ancient mine containing pockets of galena in a narrow lode which was worked in several past periods. After lying idle for many years it was reopened in 1806 and, fifteen years later, was yielding 400 tons of lead a year, a figure which increased later to 540 tons. It was drained down to the 110 fathom level by steam and water power only to close down finally in 1877.

The famous Bere Alston mines in the small peninsula at the confluence of the rivers Tamar and Tavy have been briefly described in the chapter on silver. The last workings of these mines since 1848 when Mineral Statistics began to appear gave :

Tamar Mines 1848 to 1863	13,291 tons lead ore
Tamar Consols 1849 to 1860	7,116 tons lead ore
East Tamar Consols 1848 to 1861	2,338 tons lead ore

Of course a much greater tonnage was produced in the sparsely recorded output for the previous centuries from the time of Edward I.

There was a group of lead mines in the upper Teign valley, south Devon. The Frank Mills, Wheal Exmouth and Wheal Adams mines were all on the same lode which is traceable for miles along the valley; Frank Mills was the deepest and most important of these three, producing 14,303 tons of lead ore from 1857 to 1880.

In east Cornwall there is a crosscourse running north to south across Hingston Downs, three miles west of the Tamar, but it never produced large quantities of ore. A mile further west and roughly parallel with this crosscourse is the Holmbush vein which passes through the town of Callington. It was worked to below the 150 fathom level. Four miles still further west, the Mary Ann—Trelawny lode was worked from 1843 for three miles along its course. Like all other lead veins in the west of England it was a true fissure lode. Horses of barren rock and huge

17 Macnab stands aghast as water from the river Alyn pours out of a timber shute in a lode at Halkyn District United Mines, Wales.

18 A natural limestone cavern at Powell's Lode, Halkyn. Note the rakes of one-ton mine cars.

19 An old stone-lined shaft only a few feet in diameter on The Derbyshire Rakes.

20 Dirtlow Rake, Pindale Mine, Derbyshire. Note the narrowness of workings to extract all the galena with the minimum of gangue (country rock) in these old workings.

Driving the sea level tunnel face
limestone at Halkyn District United
ines. This advance was claimed as
European tunnelling record.

22 Underground railway in the sea level tunnel, H.D.U.M. Collecting and distributing miners over six miles of workings.

23 Ruined buildings of an eighteenth-century lead mining complex, Old Gang Smelt Mills, Swaledale, Yorkshire. The 391ft long peat-house is on the skyline to the left.

24 Grinton Smelt Mill, Swaledale. The only surviving roofed mill of the Yorkshire Pennines lead-smelting industry. The building in the middle distance is the fuel store and in the background, ascending the hillside, is the partly collapsed 1,100ft long flue.

25 Underground pumping station. A Mather & Platt installation, in the below sea-level tunnel at Powell's Lode, H.D.U.M., then the largest underground plant in Britain. This shows the water problem encountered in all limestone country necessitating large pumping installations and heavy costs.

26 1000ft compressor underground at Lode 675, H.D.U.M., This was probably the first compressor to be operated underground in a British metal mine.

27 A crushing mill of the early eighteenth century. Killhope Wheel, Upper Weardale at the west end of the Park End Lead Works.

28 Lady Isabella, the Laxey waterwheel on the Isle of Man.

29 Laporte's Cavendish Mill. The largest acid grade fluorspar producer in Europe.

30 Laporte's Sallet Hole Mine, Derbyshire. Load, haul and dump equipment able to negotiate 1 in 6 gradients.

31 Ruined engine and crushing houses of Gawton Arsenic Mine, near Tavistock, Devon.

32 A ruined flue at Gawton Mine with Rumleigh Arsenic Work stack in the middle distance.

vughs were met with, these natural open caverns were many fathoms long and over ten fathoms high. In the higher levels both carbonate and phosphate of lead were found. The galena occurred as small lenticular bodies joined by a thread of mineralization. Trelawny proved productive for a length of 500 fathoms and Mary Ann for 300. The ore realized close on £1million; royalties amounted to £100,000 and dividends paid were £130,000, these large sums being mostly due to the silver content of the ore. About a quarter of a mile west again, the Ludcott and Wheal Wrey mines sold 10,413 tons of lead ore between 1853 and 1866. Herodsfoot mine, three and a half miles south-west of Liskeard, was poor until a change of management made it prosperous. Worked cheaply, it made profits of £5,000 a year.

There are many lead veins in north Cornwall running from north to south at short intervals along the coast from Tintagel to St Agnes but they were not much explored nor were they profitable. The Guarnek mine, three miles north of Truro was worked as long ago as 1720 when some of the lead ore had 100 ounces of silver to the ton. In 1814 it was reopened for two years to produce 850 tons of dressed ore. The ore was smelted and the silver extracted at the mine gave 70 ounces of silver to the ton of lead metal.

John Taylor & Sons worked Wheal Rose with success in the early nineteenth century. East Wheal Rose's lode crossed Shepherd's lode nearly at right angles and proved highly productive for many fathoms each side of the junction. When Robert Hunt started his *Mineral Statistics* in 1843 it was the most important lead mine in the West of England. In that year it produced 5,333 tons of dressed lead ore. Up to 1855 it sold ore to the value of over £750,000 and the sale of the mine that year enabled the managers to distribute £274,500 to the lucky shareholders. The new owner, however, lost £30,000. At Wheal Chiverton, three lodes met and stopes 27 foot wide were excavated at the 90 fathom level where there was an ore shoot 1,200 feet long.

The country between Porthleven and Helston yielded both lead and silver. Wheal Pool was worked at the end of the sixteenth century and was reopned and profitably exploited in the mid-eighteenth century. In 1790, after being shut for a short time, this mine yielded ore with from 30 to 60 ounces of silver to the ton. Swanpool mine near Falmouth sold 6,022 tons of lead ore with a high silver content from 1854 to 1860.

The gravity concentration of galena, the common ore of lead, was always simple because of the considerable difference in both weight and colour from the accompanying rocks. For instance, the specific gravity of galena is from 7.2 to 7.7 whereas quartz is 2.5 to 2.8 and even heavy spar, *barytes*, is much lighter at 4.3 to 4.7. The two universal operations after crushing were buddling and jigging.

The *buddle*, sometimes called a separator, might have been anything from some gently sloping boards to a circular masonry pit with mechanically operated revolving arms or sweeps to distribute evenly the crushed ore (see Plates 17 and 18). In the primitive buddle a current of water removes the lighter waste rock or gangue leaving the heavier metallic ore, and the separation is assisted by constantly stirring with a shovel, rake or other tool. Strictly however, the buddle is a shallow vat with either a convex or concave bottom and not a platform or table. A dumb buddle is one without revolving arms and a square buddle was a shallow oblong pit in which water poured on to the crushed ore was kept stirred by the naked feet of a boy who thus agitated the ore to assist the removal of the gangue.

In medieval times jigging was effected manually by a man or boy shaking the crushed ore in an apparatus like an ordinary garden sieve in a large tub filled with water. In moving the sieve vertically up and down repeatedly the ore sank to the bottom and the waste rock collected at the top to be scraped off. All jigs consist essentially of a box filled with water and containing a second, smaller box filled with sized crushed ore supported on a sieve or screen bottom. Either this inner box is reciprocated vertically or the water is pulsated through the fixed screen. The repeated movement in both cases causes the heavier ore to sink to the bottom. The waste rock fragments either flow over the top or are scraped off. The clean concentrate on the bottom is collected and usually a middle product is set aside for retreatment.

Since the introduction of flotation and 'sink and float' techniques, the overall recovery and grade of concentrates have improved markedly. At Halkyn during the 1930s the lead concentrates were eventually over 80 per cent lead and the recovery over 95 per cent.[35] Interesting details of lead ore dressing over a century ago are given in Robert Clough's excellently illustrated book on the *Lead Smelting Mills of the Yorkshire Dales* (1969).

The oldest type of lead smelter was known as the 'bole* hill'. Its normal site was near the top of a hill facing west or south-west, in the

* Sometimes written 'bail' or 'bayle'.

direction of the prevailing wind. A low stone wall with holes in the sides to let in air, surrounded a space a few feet in diameter. In front was a hollow scooped out of the earth and lined with clay having a runoff channel leading in to it. Brushwood and peat were placed within the wall with some ore on top. When a good steady wind was blowing, the fire was lit and a good draught sufficed to create a high enough temperature to reduce the ore. As the galena, sulphide of lead, was converted in part by the oxygen of the air to oxide and sulphate, the reaction between them and the galena became self-reducing, creating lead metal and sulphurous acid fumes. Fuel, usually peat, and ore were fed from time to time and the molten lead flowed out by the channel and was collected in the clay-lined basin in front. The liquid metal was scooped up in an iron ladle and poured into moulds.

In time an artificial blast of air was substituted for the natural air current. The remains of such an early furnace, dating back to Roman times, were discovered at Pentre on the edge of Halkyn Mountain in North Wales. The foundations were made of large stone blocks set in clay and lined with clay. It had been used several times with its clay lining renewed each time. There was a great deal of slag on the site the analysis of which suggested that lime and sand had been added as fluxes during smelting to produce a fluid slag. A two-inch round piece of lead proved to be the plug of the tapping hole. It is thought that the Romans used such furnaces in Derbyshire and Yorkshire, but after their departure the natives reverted to the primitive bole hill method of smelting lead. Farey[36] lists seventeen of them in Derbyshire alone and says the wind-smelting places are easy to find from the lack of any herbage on the ancient slags and ashes.

William Humphray was a goldsmith and one of the promoters of the Company of Mineral and Battery Works in 1568 which owned lead mines in Derbyshire. In 1565 he was granted a patent for smelting lead ore with a furnace and bellows and also for separating lead ore from waste by agitating the 'bouse', broken down into small pieces, in a sieve about 18 inches in diameter with two handles and with ¼ inch mesh of brass or iron wire in a tub of water, a forerunner of the mechanically operated jig. His furnace had a hearth of four stones set up on a work-stone, the ore being stirred with a gavelock (an iron poker or lever). He claimed to save over half the fuel consumed in a 'bole'. The furnace was a true ore-hearth, the blast being provided by a pair of waterpowered bellows on which he laid great emphasis. He sought injunctions against

Figure 11 Olwyn Goch headgear and monolith at Halkyn District United Mines, north Wales

other smelters who, he claimed, had infringed his patent without licence. Humphray died in 1579 and his claims against prominent smelters such as the Earl of Shrewsbury, Henry Cavendish, Sir John Zouch and others had ended in stalemate by 1584.

The true ore-hearth was the next advance, a simple easily constructed furnace, easy to operate. It was a hearth of large stones forming a box about 2 feet square and 15 inches deep. The bellows nozzle was at the back and there was a shelf-like plate in front sloping a little outwards. Bedded down with alternate layers of peat and ore, when the fire was lit the bellows soon brought the temperature to the point of reduction and beads of lead dropped to the bottom of the hearth. Much working of the ore and cinders into the hottest part of the furnace was necessary and lime was added to make a slag liquid enough to be drawn off from above the molten lead. More fuel and ore were added until sufficient metal had formed to be drawn off. Several times a shift, the lead was drawn off into a heated sumpter pot and ladled out into pig moulds. In the early days it was an ideal furnace for small mines where peat was available for the digging and drying and coal was dear and distant.[37]

Much more sophisticated ore-hearths are described by Percy, and in the early part of the eighteenth century lead smelting in Derbyshire was conducted exclusively in ore-hearths with charcoal or dried wood (called white coal) as fuel and with bellows powered by water wheels.

Up to the middle of the nineteenth century ore-hearths of this type were in use in the North Pennines and Yorkshire by small concerns and were in some cases still in use until the mines closed down at the end of the nineteenth century.

In its later development the ore-hearth was a small rectangular blast furnace composed of various cast iron components set in brickwork, not necessarily refractory. It was contained in a chamber open in front to a certain height and connected with flues suitable for condensing the lead fumes. This was an essential arrangement for protecting the workmen from the injurious lead vapour and sulphurous fumes ascending from the hearth. The hearth box was of cast iron and the workstone a flat plate of cast iron connected to the front edge of the hearth bottom from which it sloped downwards with a raised border of about an inch on both sides and lower edge. In the middle of the plate was a groove. Component parts, though of cast iron, were called stones, as they probably were originally. The hearth bottom or box and workstone were often cast in one piece and the groove conducted the molten lead into an iron pot under the workstone.

The feed door was at the side of the furnace and through this the workman placed a peat in front of the tuyère after breaking away any slag adhering to it. The peat was cut 2½ inches square and about a foot long. The hearth was left nearly full of lead metal after the previous shift, a little coal was placed on the bottom and the rest filled with peats and broken peats. Some ignited peats were placed before the nozzle, the blast was turned on and, when all the peat was well alight a little 'browse' (the agglomerated masses of ore formed in the process of smelting from the last shift) was thrown on behind the forestone. When half of it was worked in, the lead began to flow and one of the workmen stirred up the hearth while the other, with a scraper through the feed door, removed slag from the nozzle, and then threw a peat in front of it to disperse the blast. With the lead flowing freely, ore (or bouse) was added where the fire was hottest. Stirring was repeated at frequent intervals and before each stirring a peat was put in front of the nozzle and grey slag was raked out to be retreated in the slag hearth. When the iron pot was full of lead it was skimmed and cast into pig moulds.

At the end of the shift no more ore was added; the contents were stirred up two or three times after which the blast was stopped. The browse was then removed, the slag separated from it and lead from the pot put back into the hearth bottom ready for the next shift.

In 1678 George, Viscount Grandison, obtained a patent for a method of smelting and refining lead ore in a closed or reverberatory furnace with pit coal. Eight years later he set up works near Bristol. The governors of the London Lead Company, in their minutes of 19 October 1692 decided to purchase these works from Talbot Clarke who had taken them over from Grandison, but the process did not work, so in March 1695 they closed and sold the works.

Dr Edward Wright, a physician, a metallurgist and a Quaker, perfected a true reverberatory furnace which successfully smelted lead ore with pit coal in 1690. He was a leading member of the Royal Mines Copper, who had established themselves in North Wales. Meantime, at Ryton on the Tyne a smelting works was equipped with true reverberatory furnaces, three smelting furnaces, one reducing furnace, two refining furnaces and a slag hearth.

A new smelting mill was built in 1703 at Gadlis, near Bagillt on the south bank of the Dee estuary. This also had three reverberatory furnaces, refining furnaces and a slag hearth all in excellent order. Dr Wright here improved the preparation of bone ash, bettered the construction of the test bed and successfully smelted a mixture of crushed black slag and test bed. The ore came from a number of mines on Halkyn mountain.

Wright's reverberatory furnace was horizontal with a separate fire box so that only the flame, hot air and gases bore on the ore in the furnace bed. There were no bellows, the draught being effected by a tall chimney, so no waterpower was required. Both smelting and refining was conducted in the same furnace. The fire box was a large coal-burning grate with the flames passing over a low fire bridge. The bed was a low chamber 10 feet wide at the fire end and 8 feet 6 inches at the other end, with both ends slightly curved and the bed concave, higher at the back than the front. The lowest point was near the middle of the front wall. On each side there were three small doors for rabbling. The roof was 19 inches high at the bridge, sloping in a flat arch to 13 inches at the other end. The hopper and feeding hole was placed near the middle front and the bed covered with grey slag was heated until it turned pasty and then moulded to the floor to a depth of from 6 to 12 inches. The charge was 21 cwt of ore, evenly spread, and when barely red hot, turned and stirred with the doors partly opened to effect calcining in about two hours. After this stage the temperature was raised until the charge was semi-liquid when it was pushed with rakes near to the bridge and quickly melted. The liquid lead ran to the lowest point where the tapping hole had the plug removed. The metal was caught in an iron kettle, kept warm, skimmed, and ladled into pig moulds. The slag was stiffened with lime and when sufficiently pasty raked out through the middle back door. This reduction process took about five hours with two smelters and a labourer working two cycles per shift. The charge of 21 cwt of ore containing 75 to 80 per cent lead was reduced by 12 to 16 cwt of coal to yield 14½ cwt of metal or 91 per cent of the lead in the ore, the other 9 per cent being in the slag and in the fume that collected in the flues. There was no break in tapping and charging so the process was practically continuous. Such Flintshire furnaces, according to Farey writing in 1811, were introduced into Derbyshire by the 'Quaker' Company in 1745, the first at Kelstedge in Ashover.[31] Their overall size, with two rectangular flues, was 15 feet by 11 feet, the outer brickwork lined with firebrick and the whole held together with tierods. A reverberatory furnace was most suitable for finely divided concentrates without agglomeration, lead scrap, and byproducts low in impurities.

According to Percy, the reverberatory furnace is more satisfactory than the blast furnace for the treatment of lead ores as the contact may be prolonged, the temperature controlled, and perfect intermixing accomplished by rabbling. In the blast furnace, on the other hand, any portion of the charge that becomes liquid trickles down into the hearth and is withdrawn from any further chemical action. However, the throughput of a blast furnace is large, the reaction essentially deoxidizing and the slag usually lead-bearing. Many blast furnaces were operating in Britain during the nineteenth century treating low-grade and impure lead ores profitably with coke as a fuel.

The reduction of lead in blast furnaces was introduced to Britain principally from Germany. Such furnaces were large, costly to construct and required a plentiful and constant supply of ore. Usually the ore had first to be roasted, sintered or calcined in special plant to reduce the sulphur content and agglomerate the fines.

A typical furnace was cylindrical, the shaft about 5 feet 6 inches in diameter at the tuyères and 7 feet 6 inches at the mouth top. The height from tuyères to mouth was 20 feet or more, and the hearth 2 feet deep. The tuyères were in water-jacketed iron sections and through them the compressed-air blast, preferably preheated, was introduced. The furnace lining was of refractory material. In some cases the overall height was as much as 70 feet. A typical charge would have been 3 tons of ore, one ton of matte, 6 tons of slag with 1½ tons of coke.

Given high grade galena free from impurities, the Scotch hearth was the oldest and simplest. This consisted of a shallow hearth on which a coke fire floated on a bath of molten lead. The fire was maintained by tuyères at the back above the level of the lead. Dressed ore was fed on with continuous stirring, the reduced galena sinking into the molten metal. The slag was skimmed off to be treated in the blast furnace which was used to smelt lower grade ores, high grade ores with impurities, slags and other byproducts. Fines were sintered in various forms of sintering pots. The ore, fluxes and coke were fed continuously at the top of the furnace and lead, speiss and slag drawn off at the bottom.

Refining was conducted in a pot furnace or lead softening furnace which was essentially a deep reverberatory bath. A modern softening furnace 33 feet long, 16 feet wide and 4 feet deep made of ⅜-inch riveted steel plate water-jacketed at sides and ends with a capacity of 320 tons of lead metal is installed at the Britannia Lead refinery at Northfleet, Kent. In twenty-four hours 210 tons of lead are melted, softened and tapped from this furnace.[38]

Because of the dangers of lead poisoning, and so that the smelter men should not suffer from the sulphurous acid fumes, enormously long horizontal flues terminating in a high chimney were used in the Northern Pennine smelters. Those at Allenhead and Rookhope were 3,424 and 2,548 yards long respectively. With flues at the Allendale smelt mill 4,431 and 4,338 yards, these made the impressive total of some eight and half miles of flues. They collected all the lead fume and also reduced the danger to cattle and crops with the avoidance of consequent compensation.[39] A considerable quantity of lead, which would otherwise be wastefully dispersed, was periodically collected from these miles of subterranean flues. Memorials of bygone industry, their purpose is now a mystery to many who walk the Pennine moors.

CHAPTER SEVEN

Zinc

The first zinc mineral to be exploited in this country was calamine, the carbonate of zinc. The word calamus, meaning a reed, is linked with the Latin *Lapis Calamaris* of Pliny which included the carbonate and the hydrous silicate of zinc described by James Smithson in 1803. Its name derives from the reed-like masses formed on the bottom of the furnace during smelting. In England the names were reversed, calamine being applied to the carbonate and Smithsonite to the silicate.

Calamine was discovered in Britain late in the sixteenth century as the result of a prospecting campaign sponsored by the government of Elizabeth I to provide brass to make cannon needed to fight the Spaniards—an early Defence Materials Procurement Agency. The Mendips were then the principal source conveniently near Bristol where the brass was made. This secondary activity of the Mendip Mines was to be killed about 1825 by free trade, which allowed the importation of zinc metal from the continent, though calamine was still mined on Mendip up to 1856. It was also found and mined at Wheal Mary in Cornwall, near Matlock, on Alston Moor, at Roughtongill in Lakeland, near Holywell in north Wales and at Leadhills in Scotland. Calamine is not the most important source of zinc, but it was exploited centuries before the commoner mineral blende, sulphide of zinc, ZnS, was won from the lead mines of Britain.

Sphalerite or zinc blende, the sulphide of zinc, called by the miners 'Black Jack', is the commonest of zinc minerals. The name 'blende' derives from the German blenden, to delude, from its supposed resemblance to galena although containing no lead.[1] Until the middle of the nineteenth century this 'Black Jack' was a waste product of lead mines, to be thrown away on the dumps or left on the walls of lead mine stopes. It was the last of the base metal ores to be exploited not much more than

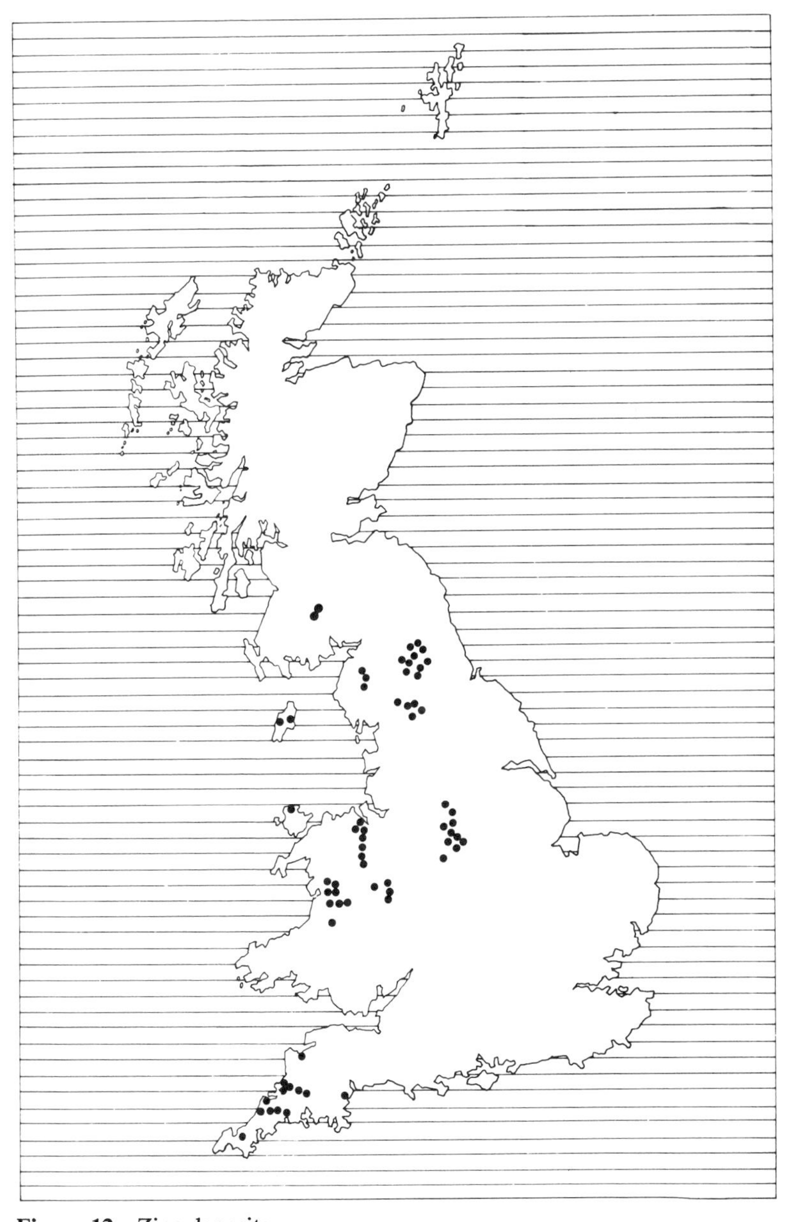

Figure 12 Zinc deposits

a century ago and the industrial archaeology of zinc is indissolubly linked with that of lead mining.

The first recorded production of blende was in 1856 when 9,003 tons of zinc concentrates were included in the United Kingdom Mineral Statistics. This chief of all zinc ores was mined from that date forward, although much was recovered from working over old lead mine dumps. Usually as the lead mines deepened the proportion of zinc blende increased.

The highest output was in 1881 when 35,527 tons of concentrates were produced,[2] after which there was a steady decline to 9,025 tons in 1918. For nearly a century from 1856 to 1949 the total output from various lead mining districts in the United Kingdom was:[3]

	Tons zinc concentrates
North Pennines	267,000
Derbyshire	80,000
Flint and Denbighshire	342,000
Cardigan and Montgomeryshire	151,000
Isle of Man	256,000
Cornwall	85,000
Shropshire	20,000
Lake District	38,000
Devonshire	4,000
Scotland, Leadhills and Wanlockhead	16,300
Total	**1,249,300** *tons*

Another way of classifying the output of zinc concentrates in Great Britain is from the various geological horizons. On this basis total production from 1856 to 1938 was:[4]

Strata	*District*	*Output of concentrates in tons*
Upper Palaeozoic carboniferous limestones and millstone grits	North Pennines	248,145
	South Pennines	62,190
	Flintshire and Denbighshire	341,601
		651,936

Strata	*District*	*Output of concentrates in tons*
Amorican	Devon and Cornwall	88,806
Lower Palaeozoic	Scotland	16,283
Slatey rocks of	Lake District	39,483
Cambrian, Ordovician	Isle of Man	255,632
and Silurian age	Shropshire	20,269
and the Caledonian	Central Wales	183,576
Granite of the Isle of	S. Wales	934
Man	Anglesey	18,419
		534,596
Trias	Cheshire	294
	Glamorgan	236
Total		**1,275,868**

From this table it will be seen that the Upper and Lower Palaeozoic rocks yielded roughly the same amounts and most of the total. Regarding Devon and Cornwall, the intense mountain building movements, granite intrusion and metamorphism gave an environment for ore deposition unparalled in this island.

The production of zinc concentrates in the North Pennines was far from simple and was produced by gravity concentration. Although the blende (specific gravity 4.0) was coarse grained it was associated in some mines with siderite (carbonate of iron Sp. Gr. 3.7 to 3.9) and ankerite (a complex carbonate of lime magnesia and iron Sp. Gr. 3.0). This occurred at the former main zinc mining centres of Nenthead and Coalclough while at Nentsberry it was associated with witherite (carbonate of baryta Sp. Gr. 4.2) and at Willyhole with barytes (sulphate of baryta Sp. Gr. 4.5). This association with heavy minerals made the production of a high grade zinc concentrate by gravity extremely difficult. About 1881 the London Lead Company produced 1,464 tons of zinc concentrates which only yielded 150 tons of zinc metal. Improvements were made by the Nenthead and Tynedale Zinc Co. which in 1889 produced concentrates averaging 40 per cent zinc, improving by 1890 to 42 per cent.

The Vieille Montagne Zinc Co. erected an elaborate new gravity plant at Nenthead in 1908. Between 1896 and 1919, this company crushed 817,574 tons of ore to produce 156,584 tons of 42.5 per cent zinc concentrates, which yielded 66,547 tons of metal.[5] The waste dumps from this operation contained 3 per cent zinc and were processed during World War II. The Vieille Montagne Zinc Co. was the principal contributor to zinc production in Britain for many years but the concentrates were smelted in Belgium. In 1913 this company produced 11,244 tons of concentrates out of 17,294 for the whole country, nearly 70 per cent of the total, but Great Laxey mine on the Isle of Man for years headed the list for an individual mine. The memory of this premier zinc ore mine is kept alive by the recently restored 74-feet diameter waterwheel—Lady Isabella (see Plate 8), conspicuously sited along the coast road. Other big producers in the past were Frongoch, Minera and Talarcoch in Wales. All these mines had closed down, however, by 1918.[6]

In an area of 300 square miles around Plymlimon in Central Wales between 1830 and 1880 over 130 mines produced between them some 180,000 tons of zinc blende. J. Taylor and Sons were active in this field for many years but abandoned their Cardiganshire ventures between 1878 and 1890 due to low metal prices. Here, as elsewhere, zinc ores predominated over lead ores in the deeper levels.

Between 1862 and 1886 the West Chiverton mine in Central Cornwall produced 22,406 tons of blende which sold for £67,217.[7] East Wheal Rose in the same area during 1882-83 sold 231 tons of blende. It was at this mine in 1843 that forty men were drowned when a cloudburst flooded the valley and the water found its way down the shaft.

As the tables show, zinc blende was produced in practically all the leadmining fields but for about ten years after World War I zinc concentrates were almost unsaleable, so they were stocked at the mines awaiting an improvement in metal prices. The low price offered was apart from low world metal prices, due to the high returning charges demanded by the smelters and the government's purchase of the output of the Broken Hill mines in Australia.[8]

In 1881 the price of spelter, the usual name for zinc metal with a little lead, was £16 13*s* but zinc concentrates only fetched £3 2*s* a ton. The metal recovery was 42 per cent. In 1913 spelter price was £22 14*s* 3*d* and the concentrate worth £4 1*s* a ton. In 1916 the price of spelter had risen to £68 8*s* 11*d* but the concentrate only to £7 12*s* 5*d*. So the mine owners complained that the price offered for the concentrates had not risen in proportion to the rise in metal price.[9] At the time one ton of lead concentrates was worth 3 tons of zinc concentrates. Before differential flotation was developed in Australia in 1922, zinc concentrates produced by gravity concentration were less than 50 per cent zinc and

ncurred high returning charges by the smelters, and it was not until
elective flotation was generally adopted that concentrates of 55 to 60
er cent zinc became the normal feed to the smelters.

When the Swansea copper smelters went out of business they took to
melting zinc. They failed, however, to keep abreast of progress and the
nechanical operation of a hot, dusty, unpleasant and fume-laden task
vas attempted too late. Byproducts like sulphuric acid were not
ecovered but added to the devastation of the surrounding landscape
nd a policy of close secrecy over technical methods was mistakenly
ollowed. An opportunity was given them to rehabilitate their works,
ut all but one failed to profit by it. All the others were out of business
y the end of World War I. The one that survived grew into a great
ompany[10] and made great progress in both lead and zinc metallurgy,
ulminating in the zinc-lead blast furnace method at the modern install-
tion at Avonmouth where their research teams have made up for the
eglect of metallurgical progress in Britain since 1914.

Although Dr Watson, a doctor of medicine of Swansea, claimed
o have been the first to smelt zinc in Europe, it was Dr Isaac Lawson
vho introduced the art into Elgland in 1730 at the Champion works at
Varmely near Bristol[11] and the secret of his process was preserved
ntil almost the end of the eighteenth century. The method 'distillation
er desensum' was to mix calamine and carbon in an enclosed fireclay
rucible and heat it to a high temperature when the zinc vapour was
listilled from a hole in the bottom of the pot by way of a protruding iron
ipe and the liquid metal caught in a suitable receptacle.

The zinc so produced at the works of William Champion came from
hirty-one furnaces. Which were at work at Warmley by 1760 and produ-
ing an average of 200 tons of metal a year, but it was not until 1781 that
inc came into general use. The calcined ore was mixed with carbon and
eated in large vessels until the oxygen burnt off as carbonic oxide gas,
nd the metallic zinc volatilized and passed through tubes attached to
he crucible. This vapour condensed into liquid form in suitable vessels
vhere it was allowed to solidify. It was then remelted and cast into bars.

In 1806, at Liège in Belgium, the horizontal retort process ousted the
English method. The concentrates were roasted or sintered for the
emoval of the sulphur. The metallic zinc was then distilled and collected
n liquid form in condensers. The metal cast into suitable moulds usual-
y contained about 2 per cent lead. The sulphide ore was roasted on an

endless moving track carrying the charge on pallets. Ignition took place at the upper front of the machine, the flame being drawn into the charge by fan suction. Because the sintered produce was cellular there was a better metal recovery in the retorts. The horizontal type of retort was from 4 feet 6 inches to 5 feet 6 inches long and oval in section (11 inches by 8 inches) and at the end of the retort was the condenser as a prolongation thereof. Because heat is applied externally, the retorts have to have thin walls but they must be capable of withstanding a high temperature without distortion. Thus they must also have great durability, conductivity, mechanical strength, tenacity and high density. Because zinc oxide and silica form slags, the retort is eventually destroyed. The furnace in which the retorts are heated must also resist severe and rapid changes of temperature of as much as 700°C. The horizontal retorts have to be limited in size[12] and one furnace may hold as many as 200 to 400 retorts. About two-thirds of the zinc content of the ore is obtained as spelter and 20 per cent as blue powder, that is, solid zinc encrusted with zinc oxide which is ground and mixed with subsequent charges to be redistilled. The charges in the individual retorts are externally heated to a very high temperature for a long time to obtain an extraction of over 80 per cent of the metal in the charges. This is a batch process which takes twenty-four hours including the charging and discharging of the retorts. The operation is onerous, unpleasant and a health hazard but is still in use. The refractories used in zinc smelting are fireclay, silica and silicon carbide. In practice, the former is the only refractory used for horizontal retorts.

The next advance was the vertical retort. This is a much larger vessel, a simple rectangular shaft 30 feet high and 7 feet by 1 foot in section, constructed of individual silicon carbide bricks whose thermal conductivity is ten times that of fireclay. Thus there is a good flow of heat from the furnace to the charge inside the retort. The zinc concentrates are dead-roasted to zinc oxide in a multiple hearth type of furnace with eight hearths and a top drier, but the middle hearths are omitted leaving two top and the two bottom beds. The roasted ore is formed into briquettes which, when heated, provide voids through which the zinc vapour can seep. There is a continuous discharge of spent briquettes at the bottom of the furnace. The silicon carbide, with its comparatively high heat conductivity, allows the walls of the retort to be relatively thick, while its low coefficient of expansion means a reduced sensibility to heat shock.

The briquettes are pillow-shaped, measuring 4 inches by 2⅞ inches by 2½ inches. They consist of approximately half sinter and half coal and after coking weigh about ¾lb. The zinc vapour is led through a condenser at the top of the furnace where the liquid zinc falls into a receiving bath. The combustion chamber surrounding the retort may be fired by oil, gas or solid fuel.

The electrothermic process developed in 1931 was the first successful method of producing a gas of the same composition as from retorts but in this case from a large furnace using the electrical resistance of the charge to supply the required heat. A single unit of this type produces 50 tons of zinc metal a day. The charge is preheated and, with its specially designed condenser, the method is a continuous process with a high condensation efficiency.

Before this, an entirely new process was the production of electrolytic zinc. It depends on the purity of a zinc sulphate solution. To render the zinc soluble, the ore is roasted thoroughly and leached with dilute sulphuric acid, the impurities being precipitated from the solution. In 1894 an experimental plant in Cornwall proved unsuccessful,[13] and it was not until 1916 that operating difficulties were overcome. An advantage of this process is that no fuel of any kind is needed and hot and dusty operations are avoided. The cells are wooden boxes lined with ¼-inch lead sheeting and the zinc is deposited on aluminium cathodes. It is, of course, essential that power is cheap and plentiful as current consumption is high. The recovery of cadmium as a byproduct is higher than in retort methods. Unlike the horizontal retort, this electrolytic method can be a continuous process.

Because the capital cost is similar, the process chosen depends on the relative cost of fuel and power. In the middle 1950s, 38 per cent of the world's zinc was produced by the electrolytic process which can produce a metal 99.99 fine. By comparison, the first product from the horizontal retorts is 99 per cent zinc although by fractionating a similar high grade 99.99 metal can be obtained.

A successful process for smelting zinc in a blast furnace was developed at Avonmouth by the Imperial Smelting Corporation after a quarter of a century's research and experimentation. In 1956 two furnaces were operating which produced between them 70 tons of zinc metal a day. The charge consists of sinterroasted concentrates and coke. The furnace gases containing 5 to 6 per cent zinc and 8 to 10 per cent CO_2 are brought into contact with a shower of molten lead when 89 per cent of the zinc vapour is condensed and recovered as metal. The process can be applied to mixed lead-zinc concentrates, the lead being tapped from the furnace bottom. These revolutionary British changes in zinc metallurgy have made up for the neglect since World War I.[14] The output from the Avonmouth and Swansea plants of the Imperial Smelting Corporation amounts to 70,000 tons of zinc a year with the added production of sulphuric acid. These plants produce from 3 to 4¼ per cent of world's zinc, but all unfortunately from imported zinc ores.

Plants using this great British metallurgical breakthrough by 1965 were producing 190,000 tons of slab zinc and 90,000 tons of lead bullion a year. However, now that mine owners are tending more and more to separate the zinc from the lead at the mines as concentrates, the necessarily enormous supply of mixed concentrates for this customs smelter has a doubtful future.

High grade 99.99 per cent zinc can be drawn or spun, which is not possible with spelter. Such metal is used for diecasting, wire, foil, anodes and optical glass, 99.95 zinc for wire, zinc oxide, and high grade brasses, 98.50 per cent zinc for sheet and strip metal, brass and other alloys and zinc oxide paints, 98.00 per cent zinc for galvanizing, zinc dust and for brass.

Brass is the most widely used of all non-ferrous mėtal alloys and copper and zinc are capable of forming solid solutions. The earliest recorded use of zinc in this country was for brass-making at Bristol in 1568. Alfa brass contains 64 per cent copper or over. Beta brass has less copper and is less amenable to working cold but easily wrought hot and contains 55 per cent copper and 45 per cent zinc; 70/30 brass has a ductility greater than any other and is the metal used for the making of cartridge cases, electric light bulb caps and door furniture; 60/40 brass is called Muntz or yellow metal and was used for many military components. 85 copper to 15 zinc is known as Pinchbeck and was used to imitate gold in cheap jewellery. It was the invention of Christopher Pinchbeck, a London clockmaker, in 1732. Gilding metals with 5 to 15 per cent zinc are mainly used as sheet, strip or wire for decoration. With the addition of 2 to 3 per cent iron and 0.561 per cent manganese 60/40 brass becomes Delta metal. Brasses melt between 850° and 1,050° Centigrade.

CHAPTER EIGHT

Associated Minerals

In the tin, copper, lead and zinc mines of Britain there were other minerals, 'gangue minerals', that up to the middle of the nineteenth century, were left underground in the stopes or thrown on the mine dumps as useless waste but later became of considerable economic importance.

For example arsenic was a godsend to the copper mining industry in west Devon and Cornwall and prolonged the life of a number of mines for as long as thirty years. The best example is Devon Great Consols which, in the forty-five years after 1878, sold 72,270 tons of refined white arsenic worth £625,062. This bolstered falling profits on copper mining.[1] This company's arsenic works at Wheal Maria were the largest in the country and covered eight acres with its five calciners and three refining furnaces, there being 4,645 feet of brick flue and a tall chimney to disperse the poisonous fumes. In 1901, with 3,416 tons, the company produced 23 per cent of the world's arsenic output.

For the decade ending 1918 Cornwall and west Devon produced a total of 22,445 tons of white arsenic, of which South Crofty contributed 7,666 tons, or a little over a third. In recent years from 1950 to 1970, output has been so insignificant that it is hidden in the small tonnage of miscellaneous products in the statistics issued by the Department of Trade and Industry (Table 125) in 1971.

All the white arsenic in Britain is produced from the mineral, mispickel, arsenopyrite, or arsenical iron pyrites (FeAsS), a lustrous mineral that if struck with steel emits sparks and a strong smell of garlic.

Native arsenic as occurring in Dolcoath and other Cornish mines is tin white in a fresh fracture, quickly tarnishing to dark grey. It does not occur in economic quantities. Mispickel was found in many mines in west Devon and Cornwall and in small quantities in Brandy Gill and

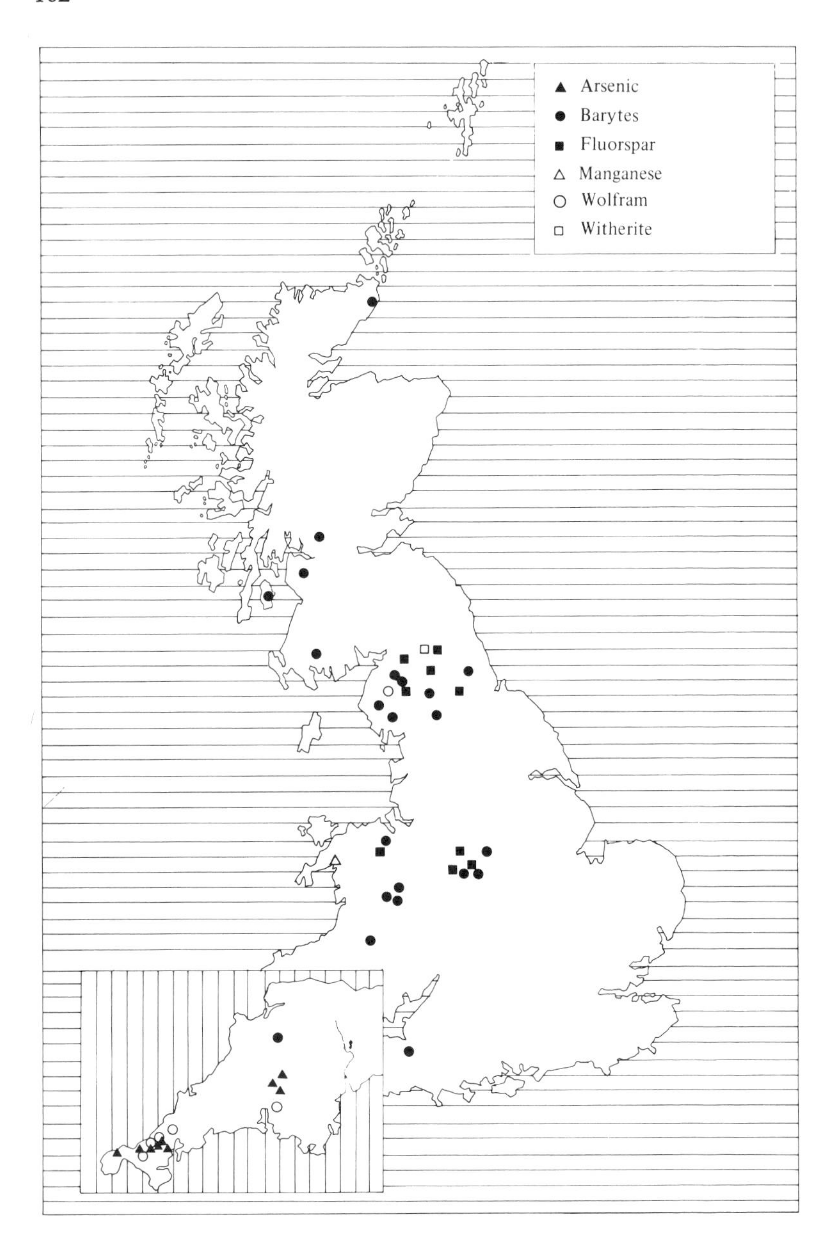

Figure 13 Associated mineral deposits

Carrock Fell in Cumberland, also at Loch Ness and Aberdeenshire in Scotland.[2] At Devon Great Consols it was said to be 5 feet thick on the stope walls after the copper ore had been extracted.

The mispickel was crushed to pass a half-inch screen, and then roasted in calciners. The earlier Brunton calciner was replaced by the 50 foot Oxland cylindrical roaster. The volatilized arsenic fumes were discharged into a labyrinth of brick chambers where they condensed as grey crystals on walls and floor. This condensate was collected by men, wearing protective clothing and ear-plugs, to be volatilized again in another set of labyrinthine flues using coke or anthracite as fuel. The resulting pure white crystals were dug out by hand and ground between two granite millstones to flour fineness, packed into blue-paper lined 4 hundredweight casks and sent down the Tamar for shipment from Plymouth. This product was 99.5 per cent Arsenious Oxide, $As_2 O_3$, and was sometimes judged by colour rather than assay.

An insect called the Boll Weevil (Anthonomus Grandis) crossed the Rio Grande from Mexico in 1893 and spread north to threaten the cotton fields of the southern United States where the quarter-inch larva lives and feeds on the inside of the cotton buds and bolls. Whenever there was a serious attack the arsenic trade throve. There was such an attack early in 1923 when a London firm of scrap metal merchants reopened Rumleigh works on the Bere Alston peninsula to produce good quality white arsenic from mispickel gleaned from mine dumps and other sources.[3] Also, at Devon Great Consols the Duke of Bedford continued to produce arsenic until 1925 from mispickel won underground from those workings that were still above water level. Until recent years the tall chimney of the works was prominent from the main road at Gulworthy. The aim of the tall chimney was to disperse the poisonous fumes and so reduce the damage to all vegetation[4] and also to bees.

The white arsenic market today is dominated by the vast tonnage stores in sealed silos of the Boliden mines in Sweden on the Baltic coast, enough to satisfy the world demand for many years to come. Besides its use in 'white' form in insecticides, crop dressings and sheep dips, arsenic is used in dying and calico printing, in copper alloys for ornamental purposes, in lead for making shot, in Paris Green, in cupric-aceto-arsenite, a pigment and insecticide, and in glasswork, medicines and paints. Where small doses occur in the drinking water supply it is said to improve the complexion to a marked degree.

Barytes, barite, heavy spar, sulphate of barium, $BaSO_4$, as it is variously called, is a common gangue mineral in British lead and zinc mines, usually associated with galena. Its specific gravity is 4.3 to 4.7. The word barium is from the Greek 'barus' meaning heavy. Before its commercial importance grew in the second half of the nineteenth century, it was thrown on the dumps or left in the stopes. During the period 1850 to 1946 a grand total of 2¼ million tons was produced from the northern Pennines, the Lake District, Shropshire and Montgomeryshire, Derbyshire, Devon and Scotland.[5] Production had increased to 90,000 tons per annum by 1939, and during World War II, when there were no imports, this tonnage was maintained. The main Scottish source, Glen Sannox on the Isle of Arran, closed before 1939.

In 1913 the domestic output was 50,045 tons,[6] but imports that year mainly from Germany were 54,630 tons because consumers in this country preferred the German product. In some old lead mines here the barytes output was more important than lead or zinc ores. Pre–World War I the price of white barytes varied from 45*s* to 60*s* a ton and off-colour from 20*s* to 25*s* but by 1920 there was no market for off-colour.

Although 80,000 tons were produced in 1950, 50,000 tons in 1960, and 24,000 tons in 1969, by 1970 the only producer was the Closehouse mine near Middleton in Teesdale, but Laportes at their Cavendish Mill in Derbyshire produce some 10,000 tons a year as a byproduct. In recent years it has become cheaper to import barytes from North Africa, where there is cheap labour and opencast workings, than to win it underground from old mines here. Much of this mineral, from Derbyshire especially, was formerly picked from old mine dumps, a procedure known as 'hillocking', but it yielded a comparatively low grade product. When cleaned, ground and bleached barytes is used as an extender in certain titanium pigments and other paints. Lithophone, a mixture of zinc sulphide and barytes, consumes about 70 per cent of the output. It also renders lead pigment more permanent. It is used as a filler in linoleum rubber, paper and textiles, and when added to materials sold by weight, increases the price as its specific gravity is about 4.5. Off-colour barytes is used in oil drilling mud and as a medium in the float and sink process of coal cleaning.[7]

Though not strictly to do with metal mining, it is worth noting here that in some collieries in south-east Northumberland and adjoining Durham the mine waters have long been known to carry barium chloride in solution. At Backworth Colliery, about ten miles north-east of Newcastle, the mine water is pumped to storage tanks. It is then filtered and mixed with sulphuric acid, when precipitated barium sulphate, or blanc fixe, is produced. In 1949 about 3,500 tons were sold by the colliery.[8]

Fluorspar gets its name from its use as a metallurgical flux from the German Fluss Spat. It is the native fluoride of calcium, fluorite, and CaF_2. It is a non-metallic mineral associated with the lead-zinc veins of the carboniferous limestone series but unlike arsenic and barytes has not suffered from foreign competition.

From 1868 to the end of the century only about 270 tons a year was produced, most being dumped or left *in situ*. Then about 1899 its value as a flux in the steel industry was realized and the market grew quickly to respectable dimensions; by 1903 nearly 12,000 tons were produced, growing to 40,000 tons by 1906. Production continued to increase to 85,000 in 1960, 115,000 in 1965 and in 1970 up to 189,000 tons, and 241,000 tons in 1971, a figure which fully meets the home demand and leaves a surplus for export.[9]

For the period 1900 to 1938 County Durham provided 624,276 tons of which 608,120 came from Weardale, largely from Sedling mine. Much of Derbyshire's 764,583 tons was lower grade material from 'hillocking'. Similarly 40,188 tons were won from Greenhow and Grassington in north Yorkshire, while insignificant quantities were produced in Devon and Cornwall.[10] During World War I, 20,154 tons out of a total output of 54,731 tons in 1916 came from the old lead-zinc mine dumps. In Derbyshire the old dumps have been picked over and over again for galena, blende, barytes and are still producing a small amount of fluorspar.

Various grades are produced: 75 to 85 per cent CaF_2 largely produced by the British Steel Corporation in the northern Pennines and goes to the steel industry where it is used in the production of ferro-silicon, ferro-manganese and ferro-vanadium alloys. Higher grades are used in ceramics, enamelware and in the non-ferrous metal industries. The highest grade, 95 per cent CaF_2 and over, is used in the chemical industry to produce hydrofluoric acid. Today the highest grades are obtained by flotation which, as far back as 1959, amounted to 14,000 tons. Laporte's Cavendish Mill at Stoney Middleton in Derbyshire is the world's biggest fluorspar unit, producing 130,000 tons of acid grade and 20,000 tons of metspar. About half of this output is exported. Weardale Lead, now a subsidiary of ICI, produces about 10,000 tons of acid grade fluorspar a year.

Specimens of fluorite display all the colours encountered in the mineral kingdom. This is due to impurities finely dispersed, while the phenomenon of fluorescence is produced by the highly dispersed colloidal condition of rare-earths. A unique occurrence of fluorspar known as Bluejohn is found in a still active mine near Castleford in Derbyshire, where there are successive seams in a small hill. This Bluejohn, a corruption of the French Bleu-Jaune, is found in massive form in colours ranging from blue through various shades of violet-red and brown to yellow. The Romans knew Bluejohn and are said to have paid fabulous prices for it. It then lay neglected and unknown until 1760 when Matthew Boulton of Birmingham began to use it to make handsome ornaments mounted in ormolu for the mantelpieces of the rich. When thoroughly impregnated with resin, much finer work became possible and delicately carved vases and goblets became the main products with prices ranging from £30 for a small bowl to £1,500 for a Matthew Boulton pastille burner.[11]

Tungsten or wolframium metal has a specific gravity of 18.77.[12] The former name was first used by Cronstedt in 1735 and is Swedish for 'heavy stone'. Wolfram, *the mineral*, was first known as an impurity in tin ores and was referred to as Lupi Spuma which Agricola translated as Wolf Froth. Cronstedt's mineral was a heavy lime-bearing substance named scheelite after K.W. Scheele who, in 1781, showed that the mineral contained a peculiar acid that he called tungstic acid. Two Spanish brothers named D'Eljujai, both chemists, were responsible for first isolating the metal tungsten but the sensational effect of alloying tungsten with steel was discovered by Robert Mushet, an Englishman, in a small works at Coleford in the Forest of Dean. Up to about 1750 wolfram was thought to be a tin mineral because its specific gravity is 7.1 to 7.9 whereas that of cassiterite tinstone is from 6.4 to 7.1. Wolfram (Fe Mn) Wo_4 is named from the Swedish 'Wolfrig' meaning 'eating' because it eats up tin as a wolf eats up sheep, so reducing the recovery of tin in smelting. Scheelite, tungstate of lime ($CaWo_4$) with a specific gravity of 5.9 is not associated with tin ores. Now tungsten is the element and wolfram the mineral.

Wolfram was separated from tinstone by the Oxland process which digests the finely divided mineral with sodium carbonate to produce soluble sodium tungstate. Later, this separation was performed by magnetic separators as it is slightly magnetic. Wolfram is almost exclusively confined to Cornwall and west Devon. It was found in a number of Cornish tin mines and at Wheal Friendship near Tavistock. Outside Devon and Cornwall, it is only found in Carrock mine, four and a half miles south of Caldbeck on the east side of Skiddaw in Cumberland where wolfram and scheelite were mined. Under the management of Anthony Wilson, in 1913 the Carrock Mine Syndicate achieved a 90 per cent recovery with a 50 tons a day treatment plant which continued to operate up to the end of World War I when the price of tungsten slumped and the mine was abandoned. At this mine the large north to south quartz vein had been known for generations, but as it yielded neither lead nor copper ore it was left severely alone until a German company worked the mine for wolfram from 1908 to 1911. The mine was reopened under the Ministry of Supply in World War II but, when the submarine menace ended and the supply position of tungsten consequently eased, it was abandoned.[13]

It was at this mine during World War II that a most interesting invention was successfully used. This was a self-contained portable apparatus for detecting scheelite by generating ultraviolet radiation of a wavelength to excite fluorescence. The scheelite exhibited visible light when exposed to this radiation in the underground workings, thus detecting the mineral which occurred in small blobs and stringers almost indistinguishable by eye from the rock with which it was associated.[14]

Because tungsten was a strategic metal, the Government aided home production in both world wars. Wheal Vincent was taken over by the Ministry of Munitions in World War I, a £5,000 loan was made to the Carrock Fell mine to complete its dressing plant and other sources of supply were similarly assisted. Meantime the Cornish tin mines produced wolfram concentrates as a byproduct, South Crofty contributing 466 tons and East Pool and Agar 444 tons to a Cornish output of 1,254 tons representing 91 per cent of the country's total but in spite of great-pressure by the Government on mine owners, production was disappointing.

At Hemerdon, about ten miles north-east of Plymouth, an attempt was made under Government auspices to work the low-grade wolfram tin deposit during the 1914-18 war, but the project was uneconomic and was suspended after hostilities ceased. The surface extent of the deposit was 4,000 feet by 450 feet. In 1939 Hemerdon Wolfram Ltd erected a small plant of 250 tons a day capacity which commenced operation in 1941 when the lease was assigned to the Ministry of Supply. The Ministry erected a big mill of 3,000 tons a day capacity as an emergency wartime measure (see Plate 29). The sampling of the orebody to a depth of 60 feet yielded a figure of 4.5 million tons containing 3.19 lb wolfram and 0.82 lb of tinstone. Importantly, there was no deterioration of values in depth.[15] From January 1942 to September 1943, 89,000 tons of ore were milled; then the big mill came into operation from October 1943 to June 1944 and treated 201,580 tons of ore, giving 181 ton of mixed wolfram–tin concentrates but recovering only 48 per cent of the sampling values. From January 1944 to June 1944 the plant ran at a rate of 2,000 tons a day milled. It closed because of the more favourable situation in wolfram supplies. The cost of production was considerably higher than had been estimated.[16]

Castle-an-Dinas mine, on a 703-feet hill eight miles east of Newquay, worked wolfram ore for fifty years. Although this lode mine was never a major producer, in 1916 it produced 78 tons 4 cwt of wolfram.

Tungsten has the highest melting point of all metals—3,387°C. It is used in special steels to produce high speed cutting tools. Ferro-tungsten is 80 per cent tungsten, 20 per cent iron and low in carbon. Its use as sodium tungstate in starch renders muslin non-inflammable; tungsten salts are used as mordants in dying and for hardening plaster of Paris. Other than as a steel alloy, its next most important use is for the fine wire filaments of electric lamps and vacuum tubes. In the form of tungsten carbide it has revolutionized rock boring in metal mines by providing detachable tool tips giving greatly increased speeds of penetration. A more recent use is for the arcing contacts of electric switchgear. As the oxide WO_3 it gives a yellow colour in glass and porcelain manufacture.[17]

Witherite, named after Dr Withering, is $BaCO_3$, barium carbonate, not mineralogically rare but the only commercial sources in the world are in Britain where the principal mine is at Settlingstone, near Fourstones, west of Hexham. This was originally a lead mine and up to 1938 had produced 16,902 tons of galena and 348,892 tons of witherite. The earliest records of the Fallowfield lead mine, also near Hexham, show that it had been worked for witherite as well as for lead, before 1850; it closed in 1912.

Small quantities occur in mines on Alston Moor and in Snailbeach mine in Shropshire. It occurs also in two Durham collieries.

The yearly output reflects the state of trade in the chemical industry, being high in 1923 (13,890 tons) and low in 1933 (5,111 tons).[18] The average production from 1939 to 1947 was about 12,000 tons. Having a specific gravity of 4.3, its gravity separation is effective.

The main use of witherite is in the production of Blanc Fixe, the precipitated sulphate used in paints, and as a filler. It is used also to produce the sulphide, chloride, hydroxide and nitrate of barium. It is used in brick-making to check white effervescence, in porcelain and glass. Other uses are in beet sugar production and, in the North of England, as a rat poison. A water-colour pigment is also made from it. The recorded output in 1969 was only 2,000 tons.

Pyrolusite, binoxide of manganese (MnO_2) occurs in many British non-ferrous mines in Cornwall, Devon, Somerset, Cumberland and Leadhills, but barely in recoverable quantities. The most important occurrence was in Caernarvonshire where, in the Lleyn peninsula, three mines near Rihw produced 197,000 tons of manganese ore between 1894 and 1945 with an average analysis of :

30-36 per cent manganese

7-10 per cent iron

18-20 per cent silica

0.3-0.5 per cent phosphorus

Because this ore is high in silica and phosphorus it was liable to penalties. In the same county a bedded deposit in ancient Cambrian rocks averaged 15 inches in thickness and occurred over a large area. It was worked discontinuously from 1892 to 1928 producing only about 44,000 tons of ore under 30 per cent manganese.[19] Up to the end of the nineteenth century the only mine with a steady output was the Chillaton and Hogstor property in Cornwall but the ore was only 20 to 30 per cent manganese. The most important use of manganese ore is for the manufacture of Spiegeleison and ferro-manganese used to toughen and harden steel.

A Postscript on Miners

Before the Industrial Revolution the skilled metal miner was his own master, haggled with the mine's manager or owner to make his 'bargain' and was free to work when and where he pleased. He was highly skilled in the sense that he sought and found the ore with all the necessary local geological knowledge that enabled him to win it with minimum effort, whereas the present underground semiskilled machinist needs a mining geologist to tell him where and what to excavate.

During the Industrial Revolution, before and after it, there were two kinds of underground workers in metal mines. One was the tributer who took a 'bargain' for payment of the dressed ore produced by him and his mates. These bargains were usually for so many shillings in the pound of the value of the ore raised. They were their own bosses in following the vein as they thought fit.

A Cornish example for a 'pitch' in Cook's Kitchen Mine was:

August ore paid 19th Sept. 1863 (6 men)

Earned 11*s* 6*d* in £	£ 51. 9*s* 9*d*
Deductions	—£ 25.18*s* 2*d*
Debt	—£ 1.18*s* 0*d*
Remainder	£ *23.13s 7d*[1]

That is, just under £4 a month each man. Deductions were for candles, explosives, sharpening drills, raising ore to grass, tramming, sampling, subsistence and a number of small items. A tributer was always a bit of a gambler. The method of letting pitches to a 'pare' of men who agreed to extract what they could for so many shillings for every £1 worth of ore they sent to surface was a quaint ceremony. In the old days, all the tributers would assemble in front of the count house where the principal mine captain would take his stand on an upstairs verandah. The clerk read out the work to be done and the number of men to be employed on it. A voice would bid 10*s* in the £. In reply the captain would propose 7*s*. Another voice would call 8*s* 6*d*. The captain would repeat the bid and throw a pebble into the air. If no other bid was made before the pebble touched the ground the last bidder secured the 'pitch' and the clerk entered his name and that of his mates in the bargain book.

Tributers played many tricks on the mine captains to make good ore look poor. As the captains had been working miners themselves they were up to all the tricks of the trade. This method of tributing meant that each crew's parcel of ore had to be handled separately, an impossibility under modern conditions.

The other class of underground worker was the 'tutworker' who took on the task of driving an adit, a horse level, a sough, sinking a shaft or a winze or raising a rise at so much a fathom's advance of an opening of agreed dimensions. He was paid for the actual ground he excavated. The harder he or they worked, the greater the advance, the more money earned, and the better pleased the management to get the work completed sooner.

The change from the subcontracting miner to the wage-paid employee was naturally evolutionary. In the chapter on lead it was seen that Thomas Sopwith in the northern Pennines forced the miners to work an eight hour shift five days a week. According to Shaw, the Germans in the Lake District did so as far back as Elizabethan times.[2] Throughout Britain the underground worker was intelligent and skilled but throughout the centuries he was poorly rewarded.

Among the medieval surface miners the Cornish 'tinners' are best documented. Here, the wage of a 'spalier', a labourer hired by a non-working adventurer in a stream works, was only about £3 a year if we are to believe William Beare, a vice-steward under Sir Francis Godolphin, and himself a veteran 'tinner', writing in 1586.[3] Carew, writing a little later, reckoned that a tinner's wage was from £4 to £6 a year. Working under tribute, that is getting a share of the value of the product, a tinner might gain a high reward one month and nothing the next, according to the bounty of nature. After deducting all working expenses and coinage, in 1697 all that remained to be shared by eighty thousand tinners was £40,338. This meant that each man received about £5 on which to support himself and his family for a year.

According to Tonkin in 1730 the best miners could earn from 20*s* to 27*s* a month—more if they worked extra shifts drawing water from the workings or raising ore. In 1773 women breaking ore by hand at Wheal Unity were getting 5*d* a day. Men on night shift watching the stamps got

1*s* a night. According to a government enquiry into the copper trade in 1798 the average monthly earnings of Cornish miners advanced fifty per cent between 1791 and 1798, from 30*s* and 42*s* to 45*s* and 68*s*. Even as late as 1838 working miners at St Just earned only 40*s* to 45*s* a month.

Offsetting these low wages, however, the cost of living was cheap. In the mid-eighteenth century at Penzance beef and mutton were 3*d* to 4*d* a pound, the best fish 1*d* a pound, mackerel 8*d* a score, butter from 5*d* to 6*d* a pound, barley 4*s* 6*d* a bushel, and coals from Wales 9*d* a bushel, though most country folk used turves (peat) at no cost at all.

Since the beginning of the Industrial Revolution the wages of the Cornish miner were pitifully low. By 1830 the average of all classes of underground workers at the Consolidated Mines was less than 70*s* a month. The average monthly wages earned by three classes of Cornish mine-workers, 1836-37 was:

	West of Penzance	*Mid-Cornwall*	*St Austell*	*Average*
Tributers	£2 7*s* 6*d*	£3 8*s* 0*d*	£2 19*s* 0*d*	£2 18*s* 2*d*
Tutworkers	£2 5*s* 0*d*	£2 17*s* 2*d*	£2 19*s* 0*d*	£2 13*s* 8*d*
Dressing Floor labourers	£2 2*s* 0*d*	£2 1*s* 0*d*	£2 5*s* 0*d*	£2 2*s* 8*d*

An extract from the *Mining Journal*, 1893, gave the following wage rates. Mine girls, Dolcoath . . . 1*s* to 1*s* 6*d* a day. Surface labourers £3 a month. Miners underground, £3 10*s* 0*d* to £5 10*s* 0*d* a month.

As recently as 1918 a government report[4] gave the following average wages per shift for metal mine workers over the whole country:

Miners	6*s*
Labourers	5*s*
Dressers	3*s* 10*d*
Smelters	8*s*
Smiths	6*s* 9
Carpenters	6*s*
Fitters	7*s* 3*d*

so no one can claim that the British metal miners were overpaid during the Industrial Revolution.

In the Lake District the English miners at Roughtongill mine in 1569 worked for only 6*d* a day, although presumably the more skilled German miners got more. At the same mine in 1858 nearly three centuries later the miners only got 3*s* a day and even in 1900, the Greenside miners got only 4*s* a shift.[5]

In North Wales, as documents prove, in 1894 adult dressing floor workers were paid 3*s* a day and boys were down to a shilling. A skilled joiner earned 14*s* a week out of which he had to pay 8*s* a week for board and lodging. Even among the enlightened mine managements of the northern Pennines in the mid-nineteenth century, a guaranteed minimum wage of 40 shillings a month was considered sufficient for a frugal family.

Throughout the centuries and right through the Industrial Revolution the miners were at the mercy of the middlemen and especially the smelters' agents. Two examples of abuses were the 'ticketing' in Redruth, where the Welsh copper smelters' agents compared tickets before the auction, and the Swansea zinc smelters with their high returning charges for dressed blende. There were no state smelters in Great Britain as in Central Europe, so the miners couldn't win.

Apart from low wages there were the poor and dangerous conditions of work. The older, more experienced, and knowledgeable miners could not climb the thousand feet or more of ladders from the lower levels at the end of the shift, so the work there became a young man's job and his life was shortened by bad ventilation, high temperatures and by breathing dust-laden air.

An investigation by Dr Richard Couch, mine surgeon during 1857 to 1859, showed that of 303 men and boys working underground at one mine the average age was 29 years 4 months; at Levant, with 206 employed, the average age was 28 years 10 months and at Ding Dong 26 years 1 month. Underground workers died young; according to the vicar of St Just writing in 1865 their active life usually ended at the age of forty.

As L. L. Price, writing in 1888 for the Statistical Society on Work and Wages in Cornish Mines, truly said, metal mining was 'an industry which seems to have escaped the disturbing influences of the Industrial Revolution'. As the ancient miners' lament put it:

Life is a place
Where we dig in a hole
To earn enough money
To buy enough bread
To get enough strength
To dig in a hole.

Gazetteer

The Geological Museum in Exhibition Road, South Kensington, not only shows specimens of British minerals but their distribution and also a fascinating series of dioramas showing the landscape and indicating the climate of this part of the world during the several geological periods from the thousands of millions of years old Cambrian to the more recent Triassic when the structures were formed in which the ores were deposited from which the metals we use were obtained.

In the British Museum are displayed inscribed Roman pigs of lead, Irish gold ornaments and bronzesmiths' hoards hidden in panic and never recovered until our time.

Pigs of Roman lead are to be seen at Chester and Newcastle museums and at least fifteen other provincial museums exhibit metal objects of all periods dug up locally.

In Camborne, Cornwall there is a small museum attached to the public library and, opposite, the more important Holman museum of mining machinery. Truro museum exhibits a prehistoric ingot of tin and wooden tools of the ancient 'tinners'.

In The Field

The high bleak moors near Leadhills not only include ruins of the silver, lead and zinc mining industry but also the many burns where skill and persistence can still reward the amateur with enough gold to make a ring (Plate 4). The postmistress of Wanlockhead is a useful source of information on the subject.

In County Durham, Alston Moor and Teesdale can be seen the stone arched entrances to horse levels as well as several of the villages, Nenthead for example, built by the London Lead Company.

The most rewarding area is the desolate moors of west Yorkshire, Arkengarthdale, Swaledale, Wensleydale, Nidderdale and Wharfedale where can still be examined the ruins of old lead dressing plants with the remains of buddles, the ruins of smelters and long flues running up the hillsides. The ruins of the famous Old Gang mine (Plate 23) are especially worth a visit, but these sites are not all approachable by car.

Farther south, the Derbyshire Rakes still stretch for miles across the bare country and ugly dumps of waste rock abound (Plates 19 and 20). The most famous Derbyshire mine, Mill Close, is now a lead works owned by Enthovens at Darley Dale.

North-east Wales shows much evidence of last century's great activity and also that of the revival in the 1930s at Halkyn. Central Wales still shows evidence in extensive dumps, shafts, adits and miles of leats of the enormous activity of earlier times which extended to four counties but was mainly centred in Cardiganshire. Devil's Bridge is a good centre to work from.

In west Wales in the neighbourhood of Dolgellau extending from Bontddu north-eastwards to Gwynfynydd north of the Mawddach estuary is the gold mining belt of Merioneth. Mine entrances, both adits and shafts, old dumps as well as the ruins of dressing plants are to be seen. Some of these mines, it is said, are going to be reopened.

In north-east Anglesey there are the opencast pits of Parys and Mona, two of the largest in Europe and the remains and reminder of a famous episode in British copper mining.

In south Wales near Pumpsaint in Carmarthenshire lies the Roman gold mine of Ogofau (Plate 1). Still to be seen and entered is the tapered adit leading to old galleries and stretches of the seven mile Roman watercourse that provided water to dress the ore are also still to be seen.

In Shropshire near Minsterley are the dumps and old workings of lead mines highly productive in Roman and medieval times.

In Cheshire at Alderley Edge (Plate 15) the more modern workings of the oldest copper mine in Britain can be examined.

There is little to see in west Devon of the famous silver mines of Bere Alston (Plates 5 and 6) or the largest copper mine in Europe near Tavistock.

There are three working tin mines in Cornwall but access is difficult. Chun Castle near St Just in the west is disappointing with its tumbled walls and staggered entrance. The buildings of the old smelter at Carvedras remain intact and the clocktower at Calenick is still to be seen and of course the many ruins of engine houses and chimneys alongside the road from Chasewater to Camborne on the backbone of Cornwall dating back to the copper mining boom of the eighteenth and nineteenth centuries.

The thousands of acres of desert near Swansea proving the devastation caused by copper and zinc smelting have been considerably rehabilitated and are no longer worth a visit.

Lastly but by no means least Lady Isabella at Great Laxey on the east coast of the Isle of Man marks the site of Britain's biggest zinc mine (Plate 28).

The Llanarmon District Mining Co., Limited.

SECRETARY'S OFFICES:
St. Werburgh's Chambers,
Chester.

Llanarmon,
Nr. Mold.

25 Sept 9 [illegible]

M. Francis Eggle M.E.

Dear Sir

William Pierce the joiner asked me when here to pay out of his wages for board & lodgings and to forward the balance to him per Ed. Ashton.

I have paid his board & lodgings = 5 weeks @ 8/- per week & will send the balance of — £1. 9. 5 with Ed. Ashton on Saturday next.

Yours Obediently
Jno Roberts

Halkyn Mines —
Nr Holywell

The Llanarmon District Mining Co., Limited.

SECRETARY'S OFFICES:
St. Werburgh's Chambers,
Chester.

Llanarmon,
Nr. Mold.
20 Sept 1894

R Francis Esqr M.E.

Dear Sir

The stuff today is better we have passed 120 wagons

We are carting from part of Garlands today & tomorrow we shall bring the stuff from there next week

We have a difficulty in keeping up steam, 3 tubes in the boiler nearest the engine and 9 tubes, in the the furthest boiler are leaking something must be done or we shall not be able to get on much longer.

I beg to remind you about –
W. Stone — the boilermaker
You said you would send him
over here on Saturday
Have you ordered
flooring boards?
Below is a list of the
men & boys required to work
the Dressing Plant –

Ed: Lloyd	3/-	per Day
Dd Edwards –	3/-	" "
Ed: Sheldon	3/-	" "
R. B. Davies	3/-	" "
Ed. Lewis	2/6	" "
Ed: Hughes –	1/3	" "
Rd. Wynne –	1/3	" "
Ed. Clemence	2/8	" "
W. Jones	2/6	
Ellis Roberts	2/6	

Your employees

Source References

Most of the facts and figures relating to the output of silver, copper, lead and zinc in this book were taken from Volumes 17, 19, 20, 23, 26 and 30 of the Special Reports on the Mineral Resources of Great Britain published by the Geological Survey between 1921 and 1925. These government publications were the result of thorough and prolonged field work by eminent geologists and detailed also the geology of the regions covered by them.

Abbreviations used in the notes

R.C.P.S.	*Royal Cornwall Philosophical Society*
R.G.S.C.	*Royal Geological Society of Cornwall*
Trans. I.M.E.	*Transactions of the Institution of Mining Engineers*
Trans. I.M.M.	*Transactions of the Institution of Mining and Metallurgy, London*

CHAPTER ONE—INTRODUCTORY

1 E. G. BOWEN, *Britain and the Western Seaways*, Thames & Hudson, 1972, p. 43; and GEORGE COFFEY, *The Bronze Age in Ireland*, Dublin, Hodges Figgis, 1913, p. 4 and 23. (Dr Oscar Montelius of Stockholm, doyen of prehistoric archaeology, placed Copper Age of Ireland mid-third millennium B.C.)

2 ANTONIO ARRIBAS, *The Iberians*, Thames & Hudson, 1964, p. 23.

3 GLYN DANIEL, *The Hungry Archaeologist in France*, Faber, 1965, p. 143.

4 JOSEPH RAFTERY, *Prehistoric Ireland*, Batsford, 1951, p. 128; and LESLIE AITCHISON, *A History of Metals*, 2 vols., Macdonald & Evans, 1960, 1, 87.

5 E. G. BOWEN, p. 59 (drawing).

6 SIR GAVIN DE BEER, 'Iktin, the Tin Port,' *The Listener*, 1 Sept. 1966, pp. 319 and 320.

7 DIODORUS, *Opera* V, 22.2.4. This passage is almost certainly Diodorus from *Timaeus*. A similar account is given by Pliny IV, 104.

8 Recent radiocarbon dating has moved the dates given here about three centuries earlier for the Beaker folk migration into the British Isles (c. 2,300 B.C.) as part of the great European population wandering. Present opinion is that the Beaker folk were Celtic speaking.
9 K. S. PAINTER, *The Severn Basin*, Cory, Adams & Mackay, 1964.
10 V. GORDON CHILDE, *The Dawn of European Civilisation*, Routledge & Kegan Paul, 1957, pp. 334 and 335.
11 SIR CYRIL FOX, *Life and Death of the Bronze Age*, Routledge & Kegan Paul, 1959, p. 183.
12 A. K. HAMILTON-JENKIN, *The Cornish Miner*, Allen & Unwin, 1927, pp. 28 and 80.
13 I. A. RICHMOND, *Roman Britain*, Pelican History of England, 1963, Vol. 1, 2nd edn. p. 119.
14 W. A. RICHARDSON, Personal letter on Roman mining in Derbyshire, Derby Technical College, 1950.
15 RICHMOND, p. 119.
16 PLINY THE ELDER, *Natural History*, Sillig's edition, 1851, Lib. XXXIV, Cap. XVII, Sect. 49, p. 102.
17 RICHMOND, p. 122.
18 *Ibid.*, p. 124.
19 HAMILTON-JENKIN, p. 29.
20 T. TUDOR, 'Northworthy', *Derbyshire Countryside*, 1934, iv, 90.
21 G. R. LEWIS, *The Stanneries*, Harvard Economic Studies III, Archibald Constable & Co, 1908.
22 *Laws of the Stanneries*, 1754, p. 46.
23 HAMILTON-JENKIN, p. 155.
24 R. N. WORTH, 'Historical Notes' concerning progress in mining skill in Devon and Cornwall, *R.C.P.S.*, 1872, p. 97.
25 A. RAISTRICK, 'The London Lead Company, 1692 to 1905', *Trans. Newcomen Soc.* xiv, 1933-34, 113-62.
26 J. A. HOLMES, 'Lead Mining in Derbyshire', *Mining Magazine*, Sept. 1962, p. 140.
27 A. H. STOKES, 'Lead and Lead mining in Derbyshire', *Trans. Chesterfield & Derbyshire Inst. of Mining, Civil and Mech. Engrs*, viii, 1880-1, p. 14.
28 HOLMES, p. 144.
29 PHILLIPS REPORT, Ministry of Munitions of War: *Report of the Advisory Committee on the Development of the Mineral Resources in the United Kingdom* (Chairman Sir Lionel Phillips), HMSO, 1918, p. 21, para 71.
30 HAMILTON-JENKIN, p. 187.
31 PHILLIPS REPORT, p. 21, para. 70.
32 UNIVERSITY COLLEGE OF SWANSEA, *The Lower Swansea Valley Project*, Longmans, 1967.

33 A. L. ROWSE, *The Cornish in America*, Macmillan, 1969, p. 4.
34 BETTERTON REPORT, *Report of the Departmental Committee appointed by the Board of Trade to investigate and report upon the non-ferrous Mining Industry* (Chairman: H. B. Betterton), HMSO, 1920.
35 E. W. O. DAWSON, 'Wartime treatment of the lead zinc dumps situated at Nenthead, Cumberland'. *Trans, I.M.M.*, lvi, 1946-47, 587-96.
36 S. J. TRUSCOTT, *Mine economics*, 2nd edn. Mining Publications, 1946, p. 2.
37 W. R. JONES, 'The nationalization of mineral rights in Great Britain', *Trans. I.M.M.*, liv, 1945, 35-74.

CHAPTER TWO — GOLD

1 W. BAINBRIDGE, *A Practical Treatise on the Law of Mines and Minerals*, London, 1841, p. 43.
2 D. S. DAVIES, 'The records of the Mines Royal and the Mineral and Battery Works', *Econ. Hist. Rev.* **6**, 1936.
3 *Ibid.*, pp. 209-13.
4 BAINBRIDGE, p. 41.
5 ROBERT ANNAN, 'Early literature of metal mining', *Trans. I.M.M.*, 70, 1960-61, p. 168.
6 S. BROOKE, *The Commentaries or reports of Edmund Plowden*, London, 1816, i, 310-39.
7 G. BERNARD HUGHES, 'Gold in the Welsh Hills', *Country Life*, 2 Oct. 1969, p. 847.
8 J. H. COLLINS, 'The precious metals in the West of England', *Royal Cornwall Geological Society*, 12, 1903, pp. 106-7, 103.
9 W. M. BEARE, Bayliff of Blackmore, British Museum, Harleian M.S.S. 1586; Folio 63, MS 6380.
10 RICHARD CAREW, *Survey of Cornwall*, London, 1602, p. 7.
11 HAMILTON-JENKIN, *The Cornish Miner*, p. 80.
12 J. MALCOLM MACLAREN, 'The occurrence of gold in Great Britain and Ireland', *Trans. I.M.E.*, 25, 1902-3, p. 440, *passim*. Much of the material in the following paragraphs is drawn from this article.
13 Collins, p. 108.
14 D. MORGAN REES, *Mines, Mills and Furnances*, HMSO, 1962, pp. 1 and 2.
15 T. R. H. NELSON, 'Gold mining in South Wales', *Mine and Quarry Engineering*, London, Jan. 1944, p. 4.
16 J. E. METCALFE, *British mining fields*, London, Inst. M. & M., 1969, p. 16.
17 NELSON, p. 59.
18 REES, p. 2.
19 HUGHES, p. 846.

20 WARINGTON W. SMYTH, *Mining and Smelting Magazine*, 1, 1862, p. 359.
21 L. H. L. HUDDART, 'St. David's Gold mine, North Wales', *Trans. I.M.M.*, 14, 1904, pp. 201, 202, 205.
22 HUDDART, p. 215.
23 HUGHES, p. 848.
24 TREVOR M. THOMAS, *The Mineral Wealth of Wales and its Exploitation*, Oliver & Boyd, 1961, p. 213.
25 ANNAN, pp. 168, 169.
26 *Ibid*, p. 169.
27 *Ibid*, p. 169.
28 J. A. S. RITSON, Personal communication, 1948.
29 A. E. TRUCKELL, Curator, Dumfries Burgh Museum, Correspondence, June 1971.
30 ALAN MCCALL, Suisgill Estate, Sutherland, Correspondence, June 1971.
31 W. A. LIVINGSTONE, *Helmsdale Village History*, Dornoch Tourist Office, Sutherland, June 1971.
32 W. H. DENNIS, *Metallurgy of the Non-Ferrous Metals*, Sir Isaac Pitman & Sons, London, 1960, pp. 605,606.
33 SIR T. K. ROSE, 'Refining gold bullion and cyanide precipitates with oxygen gas', *Trans. I.M.M.*, 14, 1904-05, p. 377.
34 R. R. KAHAN, 'Refining Gold Bullion with Chlorine Gas and Air', *Trans. I. M. M.*, 28, 1918-19, p. 35.
35 J. H. WATSON, 'Some observations on gold refining', *Trans. I. M. M.*, 68, 21 May 1959, Presidential Address, p. 478.

CHAPTER THREE—SILVER

1 MARGARET REEKS, *History of the Royal School of Mines*, published under the auspices of the R.S.M. Association, London, 1920, pp. 54, 55.
2 J. H. COLLINS, *The Precious Metals in the West of England*, Royal Institution of Cornwall, 1902, pp. 112-13.
3 FRANK RUTLEY, *Elements of Mineralogy*, 13th edn., London, Thomas Murby, 1902, p. 113.
4 COLLINS, *Precious Metals*, p. 113.
5 HERBERT C. HOOVER, *Principles of Mining*, McGraw-Hill, 1909, p. 28.
6 BENJAMIN VINCENT, *Haydn's Dictionary of Dates*, London, Edward Moxon & Co., 1871, p. 674.
7 J. H. COLLINS, 'Notes on the principal lead-bearing lodes of the West of England', *Royal Cornish Geological Society*, xii, 1903, 683-4.
8 DOUGLAS STUCKEY, *Adventurer's Slopes*, West Country Handbook No. 7. Bracknell, Berks, 1965, pp. 6, 7, 8.

9 FRANK BOOKER, *The Industrial Archaeology of the Tamar Valley*, 2nd imp. rev., David & Charles, 1971, pp. 55-66.
10 COLLINS, 'Lead-bearing lodes', pp. 689-716.
11 A. RAISTRICK, 'The London Lead Company 1692 to 1905', *Trans. Newcomen Soc.* xix, 1933-34, pp. 119-62.
12 JOHN BUTT, *The Industrial Archaeology of Scotland*, David & Charles, 1967, p. 240.
13 J. E. METCALFE, *British Mining Fields*, London, Inst. Mining and Metallurgy, 1969, p. 61.
14 A. RAISTRICK, Discussion of 'The production of galena and associated minerals in the northern Pennines' by K. C. Dunham, *Trans. I.M.M.*, liii, Jan. 1944, p. 227.
15 K. C. DUNHAM, 'The production of galena and associated minerals in the northern Pennines', *loc. cit.* p. 183, 190.
16 F. J. MONKHOUSE, 'The Greenwich Hospital Smelt Mill at Langley, Northumberland, 1768-1780', *Trans. I. M. M.* London, 1939 - 40, xlix, pp. 705-7.
17 JOHN PERCY, *Metallurgy of Lead*, Murray, 1870, pp. 179, 180.
18 AITCHISON, *A History of Metals*, i, 46.
19 S. W. SMITH, Discussion of 'A study of the Shapes and Distribution of Lead Deposits in the Pennine Limestones' by W. W. Varvill, *Trans. I.M.M.*, xlvi, 1937, 526-7.
20 VINCENT, p. 674.
21 PERCY, p. 121.
22 *Ibid*, p. 139.
23 BOOKER, pp. 62, 64.
24 PERCY, pp. 148, 151.
25 AITCHISON, ii, 466-7.

CHAPTER FOUR — TIN

1 HAMILTON-JENKIN, *The Cornish Miner*, p. 30.
2 G. RANDALL LEWIS, *The Stannaries*, Harvard Economic Studies III. Archibald Constable, 1908, p. 89.
These two books are invaluable sources for much of the information in this chapter; further reference is made only in the case of direct quotation.
3 W. G. MATON, *Western Countries*, 1796, i, 171.
4 MAJOR HENDERSON, Royal Inst. Cornwall Journal, 1917.
5 British Museum Add. MS. 29·762, folio 57-8.
6 Westwood Report, *Report of the Mineral Development Committee* (Chairman: Lord Westwood), HMSO, July 1949, p. 11, table IV.

7 *Ibid.*, p. 50, para 233.
8 *Ibid.*, p. 50, para 232.
9 E. G. BOWEN, *Britain and the Western Seaways*, Thames & Hudson, 1972, p. 44.
10 FRANK RUTLEY, *Elements of Mineralogy*, 13th edn., Thomas Murby, 1902, p. 179.
11 *Calendar of State Papers Domestic*, 1595-7, p. 373.
12 WILLIAM PRYCE, *Mineralogia Cornubiensi*, 1778, p. 133.
13 R. N. WORTH, Historical notes concerning progress of mining skill in Devon and Cornwall, *Reports of Royal Cornish Polytechnic Society*, 1872, p. 97.
14 L. SIMONIN, *Underground Life*, p. 459.
15 D. B. BARTON, *A History of Tin Mining and Smelting in Cornwall*, Truro, D. Bradford Barton, 1967, p. 258.
16 *Ibid.*, pp. 159, 173-4, 192.
17 ROBERT PEELE, *Mining Engineers' Hand Book*, Chapman & Hall, 1927, p. 1293.
18 *Ibid.*, p. 1295.
19 J. S. HALDANE, 'Miner's anaemia or ankylostomiasis', *Trans. I.M.E.*, xxv, June 1903, pp. 643-69.
20 Parliamentari Report, *Health of Cornish Miners*, HMSO, 1904, p. 31.
21 J. D. WILLSON, *Consolidated Gold Fields Ltd., Prospecting in Cornwall*, 9th Commonwealth Mining and Metallurgical Congress, 1969, Paper 17, p. 1.
22 Royal Cornwall Philosophical Society, 1872, p. 100.
23 S. J. TRUSCOTT, 'Slime treatment in Cornish frames', *Trans. I.M.M.*, xxvii, 1917-18, pp. 3-70.
24 H. S. HATFIELD, 'Dielectric mineral separation', *Trans. I.M.M.*, Bulletin No. 233. Feb. 1924.
25 D. B. BARTON, p. 152.
26 A. L. SIMON and R. O. SIMON, 'Note for tin recovery and tin dressing', *Trans. I. M. M.* xxxiv, Feb. 1925, pp. 338-45.
27 J. E. METCALFE, *British Mining Fields*, London Inst. M. & M., 1969, p. 5.
28 W. PHILLIPS, *Mineralogy*, 1826, p. 223.
29 PRYCE, p. 136.
30 W. H. DENNIS, *The Metallurgy of the Non-Ferrous Metals*, Sir Isaac Pitman and Sons, London, 1960, pp. 340, 341.
31 AITCHISON, *A History of Metals*, ii, 526.
32 SYDNEY J. JOHNSTONE and MARGARET G. JOHNSTONE, Minerals for the Chemical and Allied Industries, Chapman Hall, London, 1911, pp. 617, 618.

CHAPTER FIVE—COPPER

A. K. HAMILTON-JENKIN, *The Cornish Miner*, Allen & Unwin, 1927, is again the source for much of the information in this chapter. Two other works on which I have drawn extensively are:

FRANK BOOKER, *The Industrial Archaeology of the Tamar Valley*, 2nd impression, rev., David & Charles, 1971.
W. T. SHAW, *Mining in the Lake Counties*, Dalesman Publishing Co., Clapham, Yorks, 1970.

1 Phillips Report, p. 21.
2 Westwood Report, pp. 26-7.
3 REV. W. A. GILLIES, *In Famed Breadalbane*, The Munro Press, Perth, 1938.
4 J. E. METCALFE, *British Mining Fields*, Inst. of Mining and Metallurgy, London, 1969. p. 23.
5 Westwood Report, p. 27, para. 114.
6 REES, *Mines, Mills and Furnances*, pp. 50-1.
7 *Ibid.*, pp. 45, 47.
8 Westwood Report, p. 26, para. 114.
9 W. H. DENNIS, *The Metallurgy of the Non-Ferrous Metals*, Sir Isaac Pitman & Sons, London, 1960, pp. 77-82.

CHAPTER SIX—LEAD

1 W. R. JONES, *Minerals in Industry*, 3rd edn. Penguin (Pelican), 1955, p. 111.
2 W. W. VARVILL, 'A study of the shapes and distribution of the lead deposits in the Pennine Limestones in relation to economic mining', *Trans. I.M.M.*, xlvi, 1937, 479-89.
3 J. G. TRAILL, 'Notes on the Lower Carboniferous Limestones and toadstone at Mill Close Mine, Derbyshire', *Trans. I.M.M.*, xlix, 1939, p. 194.
4 Westwood Report, p. 40.
5 K. C. DUNHAM, 'The production of Galena and Associated Minerals in the northern Pennines', *Trans. I.M.M.*, liii, 1944., Fig. 15, Plate I, opp. p. 214, 206-7.
6 G. A. SCHNELLMANN, 'Applied geology at Halkyn, District United Mines Ltd.', *Trans. I.M.M.*, xlviii, 1939, p. 587, Table XXIII.
7 Phillips Report, 1918, p. 15, Table para 34.
8 Westwood Report, p. 41.
9 Department of Trade and Industry, *Digest of Energy Statistics*. HMSO, 1971, Table 125.

10 JAMES COFFIELD, *A Popular History of Taxation*, Longman, 1971.
11 J. F. HOLMES, 'Lead Mining in Derbyshire', *Mining Magazine*, London, Sept. 1962, p. 139.
12 A. RAISTRICK, 'The London Lead Company 1692-1905', *Trans. Newcomen Society*, xiv, 1933-34, 119-63.
13 JOHN BUTT, *The Industrial Archaeology of Scotland*, David & Charles, 1967, p. 91, 281.
14 R. A. MACKAY, 'The Leadhills—Wanlockhead Mining District', Symposium on the Future of Non-Ferrous Mining in Great Britain, London, Sept. 1958.
15 BUTT, pp. 42, 214, 211.
16 A. RAISTRICK, pp. 119-163.
17 K. C. DUNHAM, p. 182-4.
18 A. RAISTRICK, *Mines and Miners of Swaledale*, The Dalesman Publishing Co., Clapham (Yorks), 1955, p. 19, 23.
19 J. F. HOLMES, p. 142.
20 Phillips Report, p. 15, para. 34, Table.
21 W. T. SHAW, *Mining in the Lake Counties*, The Dalesman Publishing Co., 1970, p. 40 passim.
22 THOMAS PENNANT, *The History of the Parishes of Whiteford and Holywell*, London, 1796.
23 Betterton Report, 1920, p. 15, para. 61.
24 T. PENNANT, *A Tour in Wales*, London, 1773, Vol. i.
25 RICHARD WARNER, *A Second Walk through Wales*, Bath, 1798, pp. 244-50.
26 J. F. FRANCIS and J. C. ALLAN, 'Driving a mines drainage tunnel in North Wales', *Trans. I.M.M.*, xli, Jan. 1952, 302.
27 J. B. RICHARDSON, 'Attempts at sealing off river water from underground workings in North Wales', *Trans. I.M.M.*, lxiv, 1955, p. 211.
28 ABSALOM FRANCIS, *The History of the Cardiganshire Lead Mines*, Aberystwyth, 1872.
29 REES, *Mines, Mills and Furnaces*, p. 25.
30 WALTER DAVIES, *Llanymynech and District notes from the English Work*, London, 1868.
31 HAMILTON-JENKIN, *The Cornish Miner*, p. 92.
32 ROBERT HUNT, *British Mining*, 2nd edition, London, pp. 31, 36.
33 JOHN PERCY, *Metallurgy of Lead*, Murray, 1870, p. 537.
34 J. H. COLLINS, 'Notes on the principal lead bearing lodes in the West of England', *R.G.S.C.*, xii, Nov. 1902, pp. 687-716.
35 J. B. RICHARDSON, 'A revival of Lead Mining at Halkyn, North Wales', *Trans. I.M.M.*, xlvi, Dec. 1936, p. 381.
36 J. FAREY, *A General View of the Agriculture and Minerals of Derbyshire*, Macmillan, 1811, i, p. 382.

37 JOHN PERCY, pp. 216-491.
38 J. O. BETTERTON and H. P. WAGNER, Britannia Lead Refineryo Nrthfleet, Kent, *Trans. I.M.M.*, xlvi, April 1937, 706-7.
39 RAISTRICK, The London Lead Company 1692-1905, p. 119-62.

CHAPTER SEVEN — ZINC

1 FRANK RUTLEY; *Elements of Mineralogy*, 13th edn., Thomas Murby 1902, p. 175.
2 Phillips Report, p. 18, para. 47, and p. 19, Table.
3 Westwood Report, p. 40, Table XVI.
4 K. C. DUNHAM, 'The production of galena and associated minerals in the northern Pennines', *Trans. I.M.M.*, liii. 212, Table VIII, and pp. 190-1.
5 *Ibid.*, discussion (Amos Treloar), p. 237.
6 Phillips Report, p. 19, para. 51.
7 J. H. COLLINS, 'Notes on the principal lead-bearing lodes of the West of England', *R.G.S.C.*, xiv. Truro, Nov. 1902, pp. 700, 704.
8 Betterton Report, p. 15, para. 60.
9 Phillips Report, p. 19, para. 50.
10 STANLEY ROBSON, 'Mainly metallurgical', *Trans. I.M.M.*, lxiv, 1955, Presidential Address, p. 646.
11 AITCHISON, *A History of Metals*, p. 469.
12 W. H. DENNIS, *The Metallurgy of Non-ferrous Metals*, Pitman, 1960, p. 194.
13 *Ibid.*, p. 211.
14 S. W. K. MORGAN, 'The production of zinc in a blast furnace', *Trans. I.M.M.* London, lxvi, Aug. 1957, p. 553.

CHAPTER EIGHT — ASSOCIATED MINERALS

1 BOOKER, *The Industrial Archaeology of the Tamar Valley*, p. 23.
2 FRANK RUTLEY, *Elements of Mineralogy*, Thomas Murby, 1902, p. 195.
3 R. W. TOLL, 'The arsenic industry of the Tavistock district', *Mining Magazine*, London, Aug. 1953.
4 BOOKER, p. 14.
5 Westwood Report, p. 20, Table IX.
6 Phillips Report, p. 58, Appendix 10.
7 Betterton Report, p. 19.
8 Westwood Report, p. 19, para. 73.
9 Dept. of Trade and Industry, Table 125.

10 K. C. DUNHAM, The production of Galena and associated minerals in the northern Pennines, *Trans. I.M.M.* liii, 1944, 212, Table VIII.
11 CHRISTOPHER SCOTT, 'Who was Blue John?', *Daily Telegraph Magazine*, 28 May, 1971, p. 38.
12 F. RUTLEY, p. 154-5.
13 SHAW, *Mining in the Lake Counties*, p. 49.
14 PHILIP RABONE, 'A short-wave ultra-violet prospecting set for fluorescent minerals', *Trans. I.M.M.* liv, 1945, 235.
15 Westwood Report, p. 51, Table xx, and p. 54, paras 250-7.
16 JAMES CAMERON, 'The geology of Hemerdon Wolfram Mine, Devon', *Trans. I.M.M.* lxi, 1951, p. 1-2.
17 L. SANDERSON, *Tungsten*, Canadian Mining Journal, June 1939, National Business Publications, Quebec.
18 G. TRESTRAIL, 'The witherite deposit of the Settlingstone mine, Northumberland', *Trans. I.M.M.*, xl, 1930, 56-65.
19 J. E. METCALFE, *British Mining Fields*, The Institution of Mining & Metallurgy, 1969, p. 21.

A POSTSCRIPT ON MINERS

1 HAMILTON-JENKIN, *The Cornish Miner*, is again the source for much of the information in this chapter.
2 SHAW, *Mining in the Lake Counties*, p. 16.
3 WILLIAM BEARE, BM Add. MS No. 6380.
4 Phillips Report, p. 50, para. 351.
5 SHAW, p. 16.

Select Bibliography

Aitchison, Leslie. *A History of Metals*, MacDonald & Evans, 1960.

Barton, D. B. *A Historical Survey of the Mines and Mineral Railways of East Cornwall and West Devon*, 2nd edn., Truro, D. Bradford Barton, 1972.

—*A History of Tin Mining and Smelting in Cornwall*, Truro, D. Bradford Barton, 1967.

—*A History of Copper Mining in Cornwall and Devon*, 2nd edn., D. Bradford Barton, 1968.

—*A Guide to the Mines of West Cornwall*, 3rd imp., D. Bradford Barton, 1973.

—*Essays in Cornish Mining History*, 2 vols, D. Bradford Barton, 1968-71.

Booker, Frank. *Industrial Archaeology of the Tamar Valley*, Newton Abbot, David & Charles, 2nd imp. 1971.

Buchanan A., with Cossons Neil, *Industrial Archaeology of the Bristol Region* [includes the Mendip mining area], David & Charles, 1969.

Burt, Roger, ed. *Cornwall's Mines and Miners* [a collection of essays by George Henwood written in the 1850s for the *Mining Journal*], D. Bradford Barton, 1973.

—ed. *Cornish Mining, Essays in the organisation of Cornish mines and the Cornish mining economy*, David & Charles, 1969.

Butt, John. *Industrial Archaeology of Scotland*, David & Charles, 1967.

Clough, Robert T. *The Lead Smelting Mills of the Yorkshire Dales*, Leeds, Clough, 1962.

Davies-Shiel, with Marshall, J. D. *Industrial Archaeology of the Lake District*, David & Charles, 1969.

Dennis, W. H. *Metallurgy of the Non-Ferrous Metals*, Pitman, 1960.

Earl, Bryn. *Cornish Mining: The Techniques of Metal Mining in the West of England Past and Present*, D. Bradford Barton, 1970.

Hamilton-Jenkin, A. K. *The Cornish Miner*, Allen & Unwin, 1922; reprinted David & Charles, 1972.

Harris, Helen. *Industrial Archaeology of Dartmoor*, David & Charles, 1968.

—*Industrial Archaeology of the Peak District*, David & Charles, 1971.

Kirkham, Nellie. *Derbyshire Lead Mining through the Centuries*, D. Bradford Barton, 1968.

Lewis, G. R. *The Stannaries*, D. Bradford Barton, 1965.

Metcalfe, J. E. *British Mining Fields*, London, Institution of Mining & Metallurgy, 1969.

Nixon, Frank. *Industrial Archaeology of Derbyshire*, Newton Abbot, David & Charles, 1969.

Noall, Cyril. *Levant: the Mine Beneath the Sea*, D. Bradford Barton, 1971.

____*Botallack*, D. Bradford Barton, 1972.

Pennington, Robert R. *Stannary Law: A history of the mining law of Cornwall and Devon*, David & Charles, 1973.

Raistrick, A. *Mines and Miners of Swaledale*, Clapham, Yorks., The Dalesman Publishing Co., 1955.

Rees, D. Morgan. *Mines, Mills and Furnaces: an introduction to the industrial archaeology of Wales*, H.M.S.O., 1969.

Shaw, W. T. *Mining in the Lake Counties*, The Dalesman Publishing Co., 1970.

Spargo, T. *The Mines of Cornwall* (1865), reprinted, D. Bradford Barton, 1962.

Thomas, Trevor M. *The Mineral Wealth of Wales and its Exploitation*, Oliver & Boyd, 1961.

The *Transactions of the Institution of Mining and Metallurgy* naturally contain a great many references to the history and archaeology of the subject as will be apparent from the several Papers quoted individually in the Source References.

The *Transactions of the Newcomen Society* should also be mentioned. Here relevant papers include those by Rhys Jenkins (Cornwall), Nellie Kirkham (Derbyshire), Dr Arthur Raistrick (Yorkshire) and David Tew (Development of Man-engines).

Index